AF352122

The end of animal life: a start for ethical debate

Ethical and societal considerations on killing animals

edited by:
Franck L.B. Meijboom

Elsbeth N. Stassen

EAN: 9789086862603
e-EAN: 9789086868087
ISBN: 978-90-8686-260-3
e-ISBN: 978-978-90-8686-808-7
DOI: 10.3920/978-90-8686-808-7

Photo's cover:
Veeteelt (cow)
Flip Klatter, UMC Groningen (mouse)
Claire Backhouse (fish)
Fred Vloo, Photo RNW.org (horses)
Sini Merikallio (dog)

First published, 2016

Acknowledgements

The initiative for this book started as a result of a number of previous research projects funded by the Netherlands Organisation for Scientific Research (NWO) on the ethics of killing animals, the ethics of the prevention and control of animal diseases, and the ethics of livestock farming. Furthermore, the process of editing this book benefited from discussions with and support from colleagues at Wageningen University and Utrecht University. Most of all we would like to thank the authors for their valuable contributions, and Mike Jacobs at Wageningen Academic Publishers for all the effort that led to this publication. Finally, we would like to mention one author who unfortunately could not witness the publication of his chapter: prof. Richard P. Haynes (1931-2014). Richard Haynes has been a leading scholar in the field of animal ethics and played an important role as editor-in-chief of the Journal of Agricultural and Environmental Ethics. We hope in memory of him that this volume will contribute to the ethical and public debate on the end of animal life decisions.

Table of contents

1. The end of animal life: a start for ethical debate

On the role of the human-animal relationshipand the plurality of views on the value of animals

F.L.B. Meijboom[1] and E.N. Stassen[2]

[1]Utrecht University, Ethics Institute, Janskerkhof 13a, 3512 BL Utrecht; the Netherlands; [2]Wageningen University, Adaptation Physiology, De Elst 1, 6708 WD Wageningen, the Netherlands; f.l.b.meijboom@uu.nl

Abstract

Making decisions about the end of animal life is common practice, yet it is not morally neutral. The end of animal life is related to many societal and ethical questions and concerns. Questions such as how long should we continue to treat an animal before killing it? Or whether it could be legitimate to kill individual animals for the welfare of the herd or for the survival of future generations. This edited volume aims to get grip on the many questions related to the end of animal life. The chapters show how the plurality of views on killing animals is related to moral presuppositions by providing an overview on the ethical views on end of life decisions. Furthermore, the book contains a number of applied studies of the ethical questions related to killing animals in various practices including livestock farming, animal experimentation, companion animals, wildlife management, and fishing and fish farming. These chapters can help students, veterinarians, scientists, policy makers and many other professionals working with animals to easily get a good overview of the issues at stake and contribute to responsible decisions with regard to the end of animal life.

Keywords: animal welfare, moral plurality, killing animals, public debate

1.1 Introduction

The end of animal life is characterized by many complex questions and concerns. Some are mainly technical by nature, but most of them have a clear ethical component. This edited volume is dedicated to these ethical dimensions of the problems and concerns that arise at the end of animal life.

The initiative for this project started in the observation that making decisions about the end of animal life maybe common in many contexts, yet it is not evaluated as normal. All animals will die eventually, but the act of killing or decisions to try to keep animals alive is valued differently. If we start with killing, it generally is considered as a

*Franck L.B. Meijboom and Elsbeth N. Stassen (ed.) **The end of animal life: a start for ethical debate***
DOI 10.3920/978-90-8686-808-7_1, © Wageningen Academic Publishers 2016

moral wrong. Since the last century this moral judgment is no longer restricted to the killing of humans, but also applies to the killing of animals. Although killing animals is often evaluated differently if compared with cases in which humans are killed, the end of animal life is no longer neutral and is subject of public debate. Discussions about killing zoo animals or stray dogs, hunting, or animal disease control are only a few examples of the many debates on killing animals that have dominated the media in Europe and beyond. However, when one zooms in on these debates, many questions pop up, such as 'Why raises the death of a single giraffe in a zoo so much media attention, while the un-sedated killing of fish hardly get public consideration?' 'Why try some pet owners to keep their animals alive at all costs, while others opt for euthanasia rather quickly?' And 'why are members of the same animal species killed on different moments in their life, with different methods and for different purposes depending on the practice they live in?' To understand and explain these differences a mere reference to the alleged ignorance of animal keepers or the general public will not suffice. The differences have a normative ethical background: we lack a standard moral evaluation of animals and there is no univocal relationship between humans and animals. To deal with this situation, a better understanding of the ethical background of killing animals is essential. This entails more than an ethical evaluation of specific killing methods or treatments to keep animals alive. With the chapters of this book, we aim to look beneath the surface of the practices in which animals are killed or in which we try to keep animals alive. We take the current practice as a start and try to trace and explicate its normative ethical background. This ethical reflection is a key to a better understanding of the public debates on killing animals and to responsible decisions at the end of animal life. Furthermore, it is an essential element for innovations in policy on and practical methods of killing animals and the ethical justification of treatments to keep animals alive.

Putting it in this way, it may come as a surprise that there is not much more literature on the ethics of killing animals. Of course discussions on killing animals are often integrated in accounts on animal ethics, but books like the work by McMahan (2002) still seems exceptions. This is not a matter of mere indifference. Decisions at the end of animal life are intrinsically difficult. Not in the last place because such decisions are complex and irreversible. Furthermore, death is a theme that still is surrounded by taboos and is not openly discussed, e.g. slaughterhouses in Europe are often not easy to find for consumers and do not actively advertise about their quality and competence. Given this background and the wide variety of questions, this edited volume will not address all problems related to the killing of animals. However, the chapters help to explicate the normative background of the debate and help to define the context and limits of the ethical questions at the end of the animal life. With this we aim to contribute to the theoretical and practical debates on the decisions at the end of animal life.

Before giving an overview of the contributions, we analyse the impact of the mentioned lack of a standard moral evaluation of animals and of a univocal relationship between humans and animals on the discussions on killing animals.

1.2 The human-animal relationship

Animals have a special position in Western society and human-animal interactions are considered to be important and valuable (cf. Herzog, 2011). However, in society there is no single relationship between humans and animals. Humans keep animals in many ways and for very different purposes. In some cases animals are treated as members of family and valued for their own sake. In other cases the emphasis is mainly on the instrumental value animals can have combined with a strict distinction between humans and animals. This plurality crosses all contexts and cannot easily be framed in terms of the distinction between pets and production animals or between kept and (semi)-wild animals. It is also reflected in language, we use various terms for what looks like to be the same act. Some animals are euthanized; others are slaughtered, culled or destroyed. If we look at some more detail, this diversity can be recognized at four levels.

First, the general evaluation of killing animals differs between practices of human-animal interaction. In some cases the death of the animal is considered as something that should be prevented at all costs. In other practices, the killing of the animal is perceived as the lesser of two evils, while in further cases the killing of the animal is necessary to reach the aim of the sector. Second, the conditions and criteria that are applicable in the case of killing animals are very diverse. For instance, in many countries not much has been regulated with regard to the moment of killing pets except that it should not come with unnecessary pain and stress. With regard to pest control even the latter is in most cases not mentioned. At the same time, killing an animal in animal research is often strictly regulated and humane endpoints that define criteria for the moment of killing the animal have to be formulated at the start of the experiment. Thirdly, the goal for which an animal is killed is often valued differently, e.g. prevention of public health is mostly considered to be a more important goal than killing animals for fur production. Finally, there is profound plurality at the level of the justification of the decisions taken at the end of animal life. On the one hand, there is discussion about when justification is necessary. With killing mammals, there currently is a clear consensus that this always should come with some kind of justification. However, the answer on the question whether a justification is needed changes over time. For instance, up until recently killing fish was hardly considered as an act that is in need of moral justification. However, because of new knowledge about the physiology of many fish species, these fish are considered to be sentient and therefore part of our moral community (Kaiser, 2012; Sneddon, 2006). On the other hand, there is a genuine debate on what counts as a legitimate justification. This also changes over time and between cultures. An example is the preventive sanitary slaughter of healthy animals in case infectious animal diseases, such as swine fever. Compared to 50 years ago, there is much more social resistance towards preventive slaughter (Mepham, 2016; Wright *et al.*, 2010,). Today society is more and more asking for the justification for this practice. People (intuitively) argue that killing healthy animals in order to limit economic drawbacks for the food production industry is not a sufficient argument for justification (e.g. Cohen *et al.*, 2007).

These four levels at which the plurality of views on killing animals become evident show the need of a systematic analysis and better understanding of the complex human-animal relationship.

1.3 The moral position of animals: genuine plurality

Above we have seen that people evaluate killing animal from different perspectives. This diversity is not only related to differences in culture or individual psychology, it also linked to three basic ethical questions. First, are animals morally important, and if so why? Second, what duties follow from the acknowledgment of the moral status of animals? And finally, how should we deal with conflicts of duties? In society and in the academics discipline of ethics there is no single answer to each of these questions.

The first question focuses on the position of animals in the moral community. Traditionally only humans were considered to possess moral status. In the last century a variety of rather different, but strong moral arguments have been formulated that stress that our membership to the moral community cannot be based on species membership (Singer 2011). Consequently, a number of criteria have been expressed that are considered to be necessary or even sufficient for having moral status, such as the capacities of flourishing or sentience, the possession of higher cognitive capacities and of autonomy (Warren 1997). In spite of the variety of criteria, most of them show that, at least in principle, non-human animals can enter the moral community. This makes them morally considerable, i.e. they are morally important for their own sake.

This leads to the second question, what duties follow from the acknowledgment of the moral status of animals? The recognition that animals can be part of the moral community does not immediately lead to conclusions about the moral significance of the animal. At a minimum, the recognition implies that the interests of animals should be taken into account in our moral decisions. Therefore, decisions at the end of animal life cannot longer be judged from the perspective of human interests only. Our moral judgment has to include the interests of the animal. This holds for the killing of animals, but also for those cases in which it is in the human interest to keep animals alive, e.g. because we are emotionally attached to our diseased pet. However, even if we acknowledge the moral importance of animals, the content and practical consequences of our duties are not yet fully clear. It leads, for instance to the question 'what harm is in the killing of animals'? Intuitively, one could say that in general killing harms the victim. However is that applicable to animals too? The answer to this question and the view on the evaluation of killing animals are not univocal. Some argue that it is morally wrong to take the life of animals for whatever reason. Others stress that the most important duty is to prevent suffering. These differences have their origin in the above-mentioned plurality of criteria that underlie the claim that animals have moral status. In order to understand and to deal with end of animal life decisions, it is important to analyse the theories of moral status.

Finally, end of animal life decisions are not about animals only. In almost all cases the problems or dilemmas occur because human interests and values play a role too. For instance, suppose we agree on whether and why a pig is morally important for its own sake and that we have direct duties to care for its welfare, moral problems still arise if this pig is to be killed at the end of an animal experiment that aims to develop a new drug. Similar problems can also occur when end of life decisions entail that we have to choose between animals, e.g. in the case of the management of wildlife where some individuals are killed in order to make the population more robust. These examples show the need of normative ethical theories that helps us deal with conflicts of duties. These theories, such as utilitarianism or rights-based approaches provide an answer to the question 'Which norms and values should be action guiding in a specific situation?' and 'How can an action be morally justified?'.

Especially the first part of this book is dedicated to the various answers that have been formulated to these three questions. The aim is not simply to confirm the plurality, but to provide a better understanding of the theoretical background of the different views on the ethical acceptability of situations in which we decide about killing animals or keeping them alive.

1.4 Animal welfare as common starting point?

Despite the above-mentioned plurality of views, in the public and ethical evaluation of killing animals, welfare plays a central role. For instance, preventing serious welfare problems is frequently used as a justification for killing a diseased animal. Furthermore, welfare often plays an important role in public discussions on when and how is to be killed, and on who is allowed to kill an animal. Public debates in Europe and the US regularly show that there is a common-sense opinion that the actual killing of animals should avoid suffering or at least minimise suffering of the individual animal. Consequently, many Western countries have laws and regulations that criminalize animal cruelty in killing practices. Often these regulations include guidance on humane killing, i.e. preconditions and quality criteria are formulated to safeguard the welfare of the animal. For instance, the European directive on animal research (EC, 2010) takes animal welfare as a central concept in the decision on how and when to kill animals in the experiment.

Nonetheless, a mere reference to animal welfare in the discussion on killing animals will not suffice. First, it is important to take the conceptual problems and questions of definition seriously. Even though, the notion of animal welfare is widespread, there is an ongoing discussion on its conceptualization and definition (e.g. Dawkins, 2008; Duncan, 2006; Haynes, 2011; Korte *et al.*, 2007). This is more than an academic debate. Choices made at the level of conceptualisation and definition have a direct impact on what is considered to be a welfare issue with regard to killing. For instance, if welfare is defined in terms of the absence of pain and stress, killing an animal will only raise questions if the method of killing comes with (the risk of) pain and stress. However, if one takes a perspective on animal welfare that includes the element of longevity, questions of animal welfare are intrinsically linked to each decision of killing an animal (Bruijnis *et al.*, 2016). Second, animal welfare is more than a biological concept: it has a clear normative component (Fraser, 2003; Ohl and Van der Staay, 2012; Stafleu *et al.*, 1996). Making animal welfare operational always comes with normative assumptions and decisions, such as a view on the moral importance of animals, the position of welfare amongst others values and ideas about how to weigh welfare against other values or interests. This implies a continuing exchange between science and ethics in order to deal with questions of animal welfare (Bovenkerk and Meijboom, 2013). At this level we are again confronted with the aforementioned plurality of views in society. Third, ethical accounts that put welfare central to their theory still struggle with questions of how to evaluate the death and killing of animals (e.g. Haynes, 2016; Rollin, 2016; Visak and Garner, 2015). Finally, killing animals often come with a moral intuition that attention to animal welfare does not fully cover the problems raised by the killing of an animal. For instance, recent public debates on the sanitary slaughter of healthy livestock in case of the control of zoonotic diseases like Q-fever and avian influenza show that the general discussion on killing animals goes beyond the issue of animal welfare (Cohen and Stassen, 2016; Cohen *et al.*, 2007; Wright *et al.*, 2010). Similarly, discussions about end of animal life decisions with pets raise many more concerns than mere welfare questions (Van Herten, 2016).

Therefore, we endorse the importance of the attention for animal welfare. Nonetheless, we have to be hesitant to use animal welfare as the 'lingua franca' to discuss questions at the end of animal life. To get grip on the discussion it is important to map and understand the diversity of views rather than striving to one overarching term. Therefore, in this volume we take end of animal life decisions broader than questions of animal welfare. A number of contributions take a broader perspective and aim to address the moral views that it is, for instance, not respectful to kill an animal, that it violates its rights or that it ignores a reverence for life.

1.5 Outline and contributions

The combination of (a) the plurality of views on the end of animal life that reflects the general diversity of views on the human-animal relationship and (b) the ethical discussions on the value of life, the moral position of animals and justificatory reasons for killing indicate a need to reflect on the assessment and justification of decisions at the end of the life of animals and form the basis of the chapters in this edited volume. This book contains four sections that each have their own scope and focus. The first part, entitled 'ethical theory and normative considerations' deals with basic questions of killing animals from a theoretical perspective. The chapter by Robert Heeger discussses the general question of the harm of killing and the value of life. It can be claimed that ending animal life is considered as morally wrong because animal life has value. Heeger deals with this question by expanding on the work of Paul Taylor and Jeff McMahan. In the Chapters 3 and 4 animal welfare plays a central role. First, Richard Haynes discusses killing animals as a welfare issue. He argues that in principle killing an animal takes its ability to enjoy the good things of life. He argues that life has mainly instrumental value to the animal, but that killing deprives the victim of its welfare. Next, Bernard Rollin starts from the position that what happens to a being '*matters to that being*, in either a negative or a positive way' is an essential criterion to speak about moral status. This so-called mattering-condition takes a broader perspective on welfare than just the prevention of pain. His account includes the concept of *telos* and entails that a violation of *telos* may have more impact than pain. Chapter 5 is dedicated to the question whether animals have a natural right to life. From a Kantian perspective, Heike Baranzke deals with the questions (a) whether there is a duty to refrain from animal killing and (b) whether there is a moral right to life for animals. She concludes that in a Kantian approach there is no animal right to life. However, she shows that it is possible to formulate a duty based on Kant's theory: killing animals is only justified if it is performed quickly and painless. In Chapter 6 Herwig Grimm and Martin Huth take a different approach and start in pragmatism. They claim that a pragmatist approach leads to a contextual and relational understanding of the moral consideration of killing animals. This contributes to the (theoretical) understanding of the plurality of views with regard to killing animals in society. Frans Stafleu is the last author in this section. His contribution takes the different evaluation of killing humans and animals central. Along the line of two examples he traces some normative theoretical and socio-psychological backgrounds that help to understand our mixed feelings with regard to killing in general and specifically our diversity of views on when and why animals should be killed.

Part II of this volume deals with the societal debates in the context of killing animals for animal disease prevention and control. In recent years many countries all around the world have been confronted with the outbreak of contagious animal diseases that had a direct impact on economy or public health. The prevention and control of such diseases have been the start of many debates on whether and how to kill animals. The first author in this part is Ben Mepham. In Chapter 8, he deals with the claim that selective killing animals often is an ethical requirement in the case of disease outbreaks. He shows that the validity of such a claim depends on many dimensions, such as the context in which the animals live and the alleged threats to human health. Next, Nina Cohen and Elsbeth Stassen present the results from a study to the public moral convictions about killing animals in the context of animal disease control in the Netherlands. They show the dynamic relationship between moral judgements and shifts in the relational value of animals and the development of new animal practices. In Chapter 10, Marielle Bruijnis takes this empirical background as a start and discusses the importance of longevity. The hypothesis is that the concept of longevity helps to support the moral intuition that premature culling of animals is a moral wrong. This is further developed by discussion of positions in the animal welfare debate and ethical theory and along the lines of two case studies.

Part III addresses ethical questions related to the end of animal life from different practices of animal use. The aim is to show that, next to some general questions that are applicable to all practices in which animals are killed, there are specific ethical problems that relate to specific practices of animal use. The first practice is animals that are used for food production. It is evident that many animals are killed to produce meat or for animal products such as eggs and dairy products. In addition, this practice raises another question that is discussed by Stef Aerts and Johan DeTavernier in Chapter 11. They focus on killing animals as a matter of collateral damage. This means the killing of animals that are not necessary or useful for the actual production of food, but appear to be a side effect of current economic realities. The authors show that the collateral character does not result in just accidental killings, but is characterised by a systematic killing of animals. This often results in intuitions that this killing is meaningless or disproportion and therefore to be evaluated differently from the killing for food production. Whether and how this is perceived as a moral problem depends on one's normative position, yet the chapter shows that most ethical theories acknowledge that killing animals as collateral damage appear to be (extremely) problematic. Chapter 12 deals with the practice of animal testing. In this context killing is often portrayed as a necessary evil. Nuno Franco and Anna Olsson discuss this claim by combining moral positions with practical and scientific considerations. Alternatives for killing such as re-use and re-homing of animals are discussed from the perspective of animal welfare and scientific arguments. They conclude that there is a potential for rehabilitating more animals than is currently the practice. The third practice of animal use in which decisions at the end of animal life play a central role is the companion animal sector (Chapter 13). Joost van Herten shows that this practice is characterised by two different problems. On the one hand, companion animals are sometimes killed too late causing unnecessary suffering. On the other hand, there are situations in which animals are killed too early, which raises question whether this deprives the animal from a natural lifespan and potential future wellbeing. In this context, the responsibilities of the animal owners and veterinary professionals play a central role.

Part IV is the final section and is entitled 'between wild and kept'. In this part two chapters present the discussion on killing animals that are less direct under our control or with whom we (mostly) have less strong human animal relationships. The question is to what extent these dimensions play a role in the ethical evaluation of killing animals. First, in Chapter 14 Bernice Bovenkerk and Victoria Braithwaite look beneath the surface and start with the fundamental question whether it is morally justified to kill fish. They show that in spite of our ambivalent position towards fish we can take the view that killing fish is harmful. However, even if one acknowledges that killing fish is a harm a number of questions are still to be discussed, such as the moral intuition that it is worse to kill a human being or mammal than a fish, because human or mammal life is in our view more valuable. Can such a view be justified? In Chapter 15 Bart Gremmen discusses ethical questions in the context of end of life decisions with (semi-)wild animals. The first question is whether it is possible to use and translate the norms we use for our companion animals or farm animals to wild animals. The chapter shows that the answer is not directly affirmative. Our interaction with wild animals raises specific ethical questions and need reference to additional concepts such as the quality of wildness. The contribution proposes a sophisticated view on the distinction between wild and kept.

1.6 Final remarks and a dynamic future

All chapters together provide an overview of the interesting, but complex ethical debate about decisions at the end of animal. With this book we do not pretend that all questions, theories and practices are covered. For instance, insight from history and human psychology can be highly relevant to understand the discussions about killing animals. Furthermore, not all practices in which we are confronted with end of life discussions have been covered. For example, pest control often raises rather specific views about the acceptability of killing animals that strongly differ from common sense views with regard to companion animals or even on the control of animal diseases. Rather than to build an encyclopedia-like account of killing animals, our aim has been to address the problems at the end of animal life from different perspectives and approaches in order to enable the reader to get better grip on these discussions and provide him or her with tools to address and evaluate other practices and accounts. In this way the book helps to get an overview of the issues at stake and contribute to responsible decisions with regard to the end of animal life. This can be relevant for a broad audience, such as bachelor and master students in veterinary medicine and animal science and postgraduate training for professionals. Furthermore, professionals working with animals, such as veterinarians, (animal) scientists, professionals in the agri-food sector and policy makers at governments and non-governmental organisations can benefit from the analyses about the end of animal life.

Finally, the debates that form the start of this book are characterised by dynamics. This will not change in the future. Current discussions in European countries about the acceptability of killing animals for fur and debates about the acceptability of keeping alive dogs with inherited diseases may extend to other practices of animal use and to other parts of our world. Discussions about major societal challenges, such as climate change and the growing global population will influence the ethical debate. Furthermore, new practices, such the use of insects for food, or the introduction of new technologies, such as novel breeding techniques will raise new questions about how to evaluate the end of animal life and to justify the decisions. This shows that questions about the end of animal life may change, but will not disappear in the future. There will remain a living and ongoing debate. We hope that this book will actively contribute to a better informed and well-argumented discussion on the ethical questions at the end of animal life.

References

Aerts, S. and De Tavernier, J., 2016. Killing animals as a matter of collateral damage. In: Meijboom, F.L.B and Stassen, E.N. (eds.) The end of animal life: a start for ethical debate. Wageningen Academic Publishers, Wageningen, the Netherlands, pp. 169-185.

Baranzke, H., 2016. Do animals have a moral right to life? Bioethical challenges to Kant's indirect duty debate and the question of animal killing. In: Meijboom, F.L.B and Stassen, E.N. (eds.) The end of animal life: a start for ethical debate. Wageningen Academic Publishers, Wageningen, the Netherlands, pp. 61-77.

Bovenkerk, B. and Braithwaite, V.A., 2016. Beneath the surface: killing of fish as a moral problem. In: Meijboom, F.L.B and Stassen, E.N. (eds.) The end of animal life: a start for ethical debate. Wageningen Academic Publishers, Wageningen, the Netherlands, pp. 227-249.

Bovenkerk, B. and Meijboom, F.L.B., 2013. Fish welfare in aquaculture: explicating the chain of interactions between science and ethics. Journal of Agricultural and Environmental Ethics 26: 41-61.

Bruijnis, M.R.N., Meijboom, F.L.B and Stassen, E.N., 2016. Premature culling of production animals; ethical questions related to killing animals in food production. In: Meijboom, F.L.B and Stassen, E.N. (eds.) The end of animal life: a start for ethical debate. Wageningen Academic Publishers, Wageningen, the Netherlands, pp. 149-165.

Cohen, N.E. and Stassen, E., 2016. Public moral convictions about animals in the Netherlands: culling healthy animals as a moral problem. In: Meijboom, F.L.B and Stassen, E.N. (eds.) The end of animal life: a start for ethical debate. Wageningen Academic Publishers, Wageningen, the Netherlands, pp. 137-147.

Cohen, N.E., Van Asseldonk, M.A.P.M. and Stassen, E.N., 2007. Social-ethical issues concerning the control strategy of animal diseases in the European Union: a survey. Journal of Agriculture and Human Values 24: 499-510.

Dawkins, M.S., 2008. The science of animal suffering. Ethology 114: 937-945.

Duncan, I.J.H., 2006. The changing concept of animal sentience. Applied Animal Behaviour Science 100: 11-19.

European Commission (EC), 2010. Directive 2010/63/EU of the European Parliament and of the Council of 22 September 2010 on the protection of animals used for scientific purposes. Official Journal of the European Union L 276, 20.10.2010: 33-79.

Franco, N.H. and Olsson, I.A.S., 2016. Killing animals as a necessary evil? The case of animal research. In: Meijboom, F.L.B and Stassen, E.N. (eds.) The end of animal life: a start for ethical debate. Wageningen Academic Publishers, Wageningen, the Netherlands, pp. 187-201.

Fraser, D., 2003. Assessing animal welfare at the farm and group level: the interplay of science and values. Animal Welfare 12: 433-443.

Gremmen, B., 2016. Will wild make a moral difference? In: Meijboom, F.L.B and Stassen, E.N. (eds.) The end of animal life: a start for ethical debate. Wageningen Academic Publishers, Wageningen, the Netherlands, pp. 251-263.

Grimm, H. and Huth, M., 2016. The 'significance of killing' versus the 'death of an animal'. In: Meijboom, F.L.B and Stassen, E.N. (eds.) The end of animal life: a start for ethical debate. Wageningen Academic Publishers, Wageningen, the Netherlands, pp. 79-101.

Haynes, R.P., 2011. Competing conceptions of animals welfare and their ethical implications for the treatment of non-human animals. Acta Biotheoretica 59: 105-120.

Haynes, R.P., 2016. Killing as a welfare issue. In: Meijboom, F.L.B and Stassen, E.N. (eds.) The end of animal life: a start for ethical debate. Wageningen Academic Publishers, Wageningen, the Netherlands, pp. 41-48.

Heeger, F.R., 2016. Killing animals and the value of life. In: Meijboom, F.L.B and Stassen, E.N. (eds.) The end of animal life: a start for ethical debate. Wageningen Academic Publishers, Wageningen, the Netherlands, pp. 27-39.

Herzog, H., 2011. Some we love, some we hate, some we eat: why it's so hard to think straight about animals. HarperCollins Publishers, New York, NY, USA.

Kaiser, M., 2012. The ethics and sustainability of aquaculture. In: Kaplan, D.W. (ed.) The philosophy of food. University of California Press, Berkeley, CA, USA, pp. 233-249.

Korte, S.M., Olivier, B. and Koolhaas, J.M., 2007. A new animal welfare concept based on allostasis. Physiology and Behaviour 92: 422-428.

McMahan, J., 2002. The ethics of killing: problems on the margins of life. Oxford University Press, Oxford, UK.

Mepham, B., 2016. Morality, morbidity and mortality: an ethical analysis of culling nonhuman animals. In: Meijboom, F.L.B and Stassen, E.N. (eds.) The end of animal life: a start for ethical debate. Wageningen Academic Publishers, Wageningen, the Netherlands, pp. 117-135.

Ohl, F. and Van der Staay, F., 2012. Animal welfare: at the interface between science and society. The Veterinary Journal 192: 13-19.

Rollin, B.E., 2016. Death, telos and euthanasia. In: Meijboom, F.L.B and Stassen, E.N. (eds.) The end of animal life: a start for ethical debate. Wageningen Academic Publishers, Wageningen, the Netherlands, pp. 49-59.

Singer, P., 1979. Practical ethics. Cambridge University Press, New York, NY, USA.

Sneddon, L.U., 2006. Ethics and welfare: pain perception in fish. Bulletin of the European Association of Fish Pathologists 26: 6-10.

Stafleu, F.R., 2016. Even a cow would be killed ...: about the difference between killing (some) animals and (some) humans. In: Meijboom, F.L.B and Stassen, E.N. (eds.) The end of animal life: a start for ethical debate. Wageningen Academic Publishers, Wageningen, the Netherlands, pp.103-113.

Stafleu, F.R., Grommers, F.J. and Vorstenbosch, J., 1996. Animal welfare: evolution and erosion of a moral concept. Animal Welfare 5: 225-234.

Van Herten, J., 2016. Killing of companion animals: to be avoided at all costs? In: Meijboom, F.L.B and Stassen, E.N. (eds.) The end of animal life: a start for ethical debate. Wageningen Academic Publishers, Wageningen, the Netherlands, pp. 203-223.

Visak, T. and Garner, R. (eds.), 2015. The ethics of killing animals. Oxford University Press, New York, NY, USA.

Warren, M.A., 1997. Moral status: obligations to persons and other living things. Oxford University Press, New York, NY, USA.

Wright, N., Meijboom, F.L.B. and Sandøe, P., 2010. Thoughts on the ethics of preventing and controlling epizootic diseases. The Veterinary Journal 186: 127-128.

Part I.

Ethical theory and normative considerations

2. Killing animals and the value of life

F.R. Heeger
Utrecht University, Department of Philosophy, Ethics Institute, Janskerkhof 13A, 3512 BL
Utrecht, the Netherlands; f.r.heeger@uu.nl

Abstract

This chapter deals with the thesis that killing animals is morally wrong because their life has value. The central question asked is how we should interpret this thesis. In order to elucidate two main possibilities, I discuss two outstanding but fundamentally different investigations: Paul Taylor's biocentric defence of respect for life and Jeff McMahan's account of the wrongness of killing animals and the badness of their death. I argue that Taylor's egalitarian and solely life-centred theory creates unacceptable difficulties which McMahan's account avoids.

Keywords: respect for nature, inherent worth, badness of death, time-relative interest

2.1 Introduction

Killing animals is a widely accepted social practice aimed at benefiting human beings. For example, many billions of animals are killed each year for consumption. At the same time, there is a fairly widespread moral concern about what is done to the animals. Quite a few people cannot help thinking that bringing about the premature death of animals is in conflict with the moral duty of showing consideration for their life. If one is seriously concerned about the lot of the animals, then one will not be ready to approve of the practice to kill them for consumption. Being faced with this contrast between accepting the practice and having concern for the animals prompts us to make up our mind about what moral judgement on the killing of animals we ourselves should endorse. Should we regard the practice of killing as admissible, because we consider the moral concern about the animals mistaken, or should we think the concern justified and in consequence look at the practice as something that should be changed or be given up? What judgement we should endorse hinges on the question whether there is reason to be concerned about the animals. For what reason would killing animals be morally wrong? The present chapter deals with one important answer: killing animals is wrong because their life has value. In order to elucidate what one should understand by this answer, I will concentrate on two outstanding but fundamentally different investigations. I will in the first place give an account of the way respect for life is defended in one of the most famous biocentric (life-centred) theories, Paul Taylor's theory of environmental ethics (Taylor, 1986). This account will be followed by a criticism of Taylor's theory. After that I will discuss

*Franck L.B. Meijboom and Elsbeth N. Stassen (ed.) **The end of animal life: a start for ethical debate***
DOI 10.3920/978-90-8686-808-7_2, © Wageningen Academic Publishers 2016

Jeff McMahan's fundamentally different view of the wrongness of killing animals and the badness of their death (McMahan, 2002).

2.2 Taylor's defence of respect for life

According to Paul Taylor's biocentric theory, killing animals is morally wrong because it contravenes the moral rule of non-maleficence. This rule prohibits harmful and destructive acts done by persons. It is applicable to the killing of animals because it expresses or embodies the fundamental moral attitude of respect for nature.

Taylor is convinced that the attitude of respect for nature is an essential component of a life-centred ethic. Therefore he offers both an explanation and a justification of this attitude. Let us first deal with his explanation. Taylor focuses on what it means for a moral agent to have the attitude of respect for nature. His answer to this question contains two concepts: the idea of the good of a being, and the idea of the inherent worth of a being. The content of the first concept is this. If a moral agent has the attitude of respect for nature, then she perceives and understands all animals and other individual organisms, however dissimilar to humans they may be, as beings that have a good of their own, each pursuing its own good in its own unique way. They develop, grow, and maintain their life. They can, by successfully adapting to their environment, keep the normal biological functions of their species throughout their lifetime. They can do well in this. They can thrive and flourish. What the good of different living beings precisely consists in is dependent on what sort of beings they are, what species-specific characteristics they have. There is a great variety here, yet it is possible to come to know the characteristics of different species and to form appropriate judgements on the good of living beings.

2.2.1 Inherent worth

To have the attitude of respect for nature includes much more than to recognize living beings as having a good of their own. It also means to regard these beings as possessing inherent worth, that is, to consider them to be worthy of respect on the part of all moral agents. This is the content of Taylor's second concept. To elucidate this concept, Taylor offers a contextual definition. To declare that a living being has inherent worth is to make the following claim: A state where the good of this being is realized is better than a comparable state where it is not realized or realized to a lesser degree. Taylor adds two features to this definition. A living being's inherent worth does not depend on its being valued for its usefulness in furthering the ends or the good of other beings. Nor does its inherent worth depend on its being appreciated for arousing feelings of wonderment or admiration.

Taylor's concept of inherent worth has a morally obligating character, for the assertion that a living being has inherent worth is to be understood as entailing two moral demands: First, this being is deserving of moral concern and consideration, or, in other words, it is to be regarded as a moral subject. Second, all moral agents have a *prima facie* duty to promote or preserve the good of this being as an end in itself and for the sake of the being whose good it is. So, we may say that with the concept of inherent worth, Taylor makes the claim that the good of a living being matters morally. Moral agents have direct duties towards this being, namely to have moral consideration for it and to promote or preserve its good.

Moreover, Taylor's concept of inherent worth is an egalitarian concept. If inherent worth is attributed to any living being, then each living being is understood to have the same status as a moral subject to which duties are owed by moral agents. All are held to be deserving of equal consideration. So, there is a morally basic equality of all living beings.

2.2.2 Biocentric outlook

For a clear understanding of Taylor's theory, we should also pay attention to his justification of the attitude of respect for nature. He starts out from three statements. First, the attitude we think it appropriate to take towards living beings depends on how we conceive of them and of our relationship to them. Second, the moral significance nature has for us depends on the way we look at the whole system of nature and our role in it. Third, how we conceive of living beings, of nature as a whole, and of our role in it belongs to our belief-system or philosophical world view. Taylor claims that the belief-system on which the attitude of respect for nature depends is the biocentric outlook. It is this belief-system that provides a justification for having or taking the attitude.

The biocentric outlook has four elements. The first is the conception of human beings as members of the Earth's community of life. One identifies oneself as a member of this community and sees one's membership as providing a common bond with all the different species of living beings that have evolved. The second element is the view of nature as a system of interdependence of which humans along with all other living beings are integral parts. One sees that not only the physical conditions of the environment but also the relations to other living beings determine each living being's chances of faring well or poorly. The third element is the awareness of the reality of the lives of individual organisms. One sees each of them as a teleological (goal-oriented) centre of life, pursuing its own good. The fourth element is the rejection of the idea that human beings are inherently superior to other living beings. One commits oneself to the principle of species-impartiality: No bias in favour of some species over others is acceptable. Taylor regards this denial the key to why the biocentric outlook supports a moral agent's adopting the attitude of respect for nature. Having explained the elements of his biocentric outlook, he concludes that given this world view, the attitude of respect is the only appropriate moral attitude to take towards the natural world and its living inhabitants.

To sum up, one can say that according to Taylor's theory, the phrase that 'the life of animals has value' (see Section 2.1 Introduction) is to be interpreted as a normative statement. It should mean: Animals as living beings have inherent worth. To regard them as beings possessing inherent worth is implied by having or taking the attitude of respect for nature. To assert that they have inherent worth entails two moral demands: to have moral consideration for them and to preserve or promote their good as an end in itself and for their sake.

2.3 Difficulties with Taylor's theory

Thinking about whether or not we should agree to Taylor's defence of respect for life, we meet with two difficulties. The first is caused by Taylor's egalitarianism, the second by his view that life is the only valid criterion of a being's moral status. Let us take them in this order.

2.3.1 Egalitarianism

The first difficulty is that Taylor's egalitarianism has impractical consequences. Taylor is an egalitarian. According to him, there is a morally basic equality of all living beings. All of them are equal in having inherent worth. From the perspective of a moral agent, all entities seeking to realize the good of their own, that is all living beings, have to be regarded as equal bearers of inherent worth and are to be met with the same moral respect. To have genuine respect for a living being is to act out of concern and consideration for its own good. This is a matter of principle, not of affection or appreciation.

One may say that Taylor holds a radical biological egalitarianism. An important difficulty with this egalitarianism is that it has impractical consequences. This difficulty can be illustrated with a drastic example. A radical biological egalitarianism implies, that many ordinary and essential human activities – such as cooking, cleaning, and using disinfectants – are the moral equivalents of mass homicide. If we sought to prevent such destruction, then the health of the human population would suffer severely. If we came to see homicide as no more serious wrong than we take the destruction of micro-organisms to be, then we would be threatened by uncontrolled intra-human violence (Warren, 1997). Being faced with such strange consequences, one may wonder whether Taylor's theory has too little power to draw distinctions necessary for resolving conflicts between duties one has to different kinds of living beings.

Taylor would answer this question by pointing out two things. First, that there is a morally basic equality of all living beings does not imply that all are to be treated in the same way. To do so would be inadequate, because living beings differ as to their own good, while realizing its own good is equally important to each living being. Second, one can formulate priority principles for resolving conflicts between duties of human ethics and duties of environmental ethics. In our context two of these principles are important. One is the principle of proportionality. It distinguishes basic interests of a species from non-basic ones and prohibits us from allowing non-basic interests to override basic ones. Basic are those interests that have to be fulfilled if a living being is to remain alive. In the case of human beings, basic interests are that which rational and factually enlightened people would value as an essential part of their very existence as moral agents. The other principle is that of self-defence. This principle concerns conflicts of basic interests. It says that it is permissible for moral agents to protect themselves against dangerous or harmful organisms by destroying them. However, the principle of self-defence permits killing only when absolutely required for maintaining the very existence of moral agents. It does not provide a way out of impractical consequences like those in the example above.

2.3.2 Life as the only criterion

The second difficulty with Taylor's theory is caused by his view that life is the only valid criterion of a being's moral status, or, in other words, that an entity is to be regarded as a moral subject just in virtue of its being a living organism. Living beings are owed moral concern and consideration just because they are living beings. The difficulty with this view is that it conflicts with what we may regard as a considered conviction in moral thinking. This conviction is about our having moral duties. These duties can be of two kinds. First, we can have duties towards an entity, duties that should be performed for the sake of the entity, because its needs, interests, or well-being have moral importance in their own right. Secondly, we can have duties regarding an entity, duties not for the sake of the entity, but for other reasons, for instance, the reason that the entity is for the benefit of our fellow human beings or is vital to the ecosystem. In our criticism of Taylor, the first kind of duties is at issue. If we are to have moral duties towards an entity, then more is required than the bare fact that the entity is a living being. What these further requirements precisely consist in is an essential and disputed subject matter of ethics. In our context it is sufficient to focus on one requirement that is relatively modest and widely assented to: the requirement of mattering. It says that we have moral duties towards a living being only if our actions or omissions matter to it; and they matter to it because it is capable of being aware that the thwarting of its needs is a state to be avoided (Rollin, 1992, 1995). So, according to this requirement, the beings towards which we are to have moral duties must have some sort of awareness or sentience. That they are living beings is not enough for there to be moral duties towards them.

Similar reflections are to be found in some critical comments on Taylor's view of living beings as moral subjects. Let us look at two short and fairly representative comments made by Singer and Warren respectively. Singer takes issue with Taylor over the argument that every living thing is pursuing its own good in its own unique way, for Taylor states that once we are convinced of the argument, we can see all living beings as we see ourselves and therefore we are ready to place the same value on their existence as we do on our own. The fault Singer finds with the argument is that it uses the phrase that every living being is 'pursuing its own good' in a metaphorical way. In a literal way it is not applicable to plants because plants are not conscious and cannot engage in any intentional behaviour. We may often talk metaphorically about plants 'seeking' water or light so that they can survive, but it is possible to give a purely physical explanation of what is happening; and in the absence of consciousness, there is no good reason why we should have greater respect for the physical processes that govern their growth and decay than we have for those that govern non-living things (Singer, 1993). Warren's critical comment resembles Singer's but is slightly different. According to Warren, the fact that living beings are teleological or goal-directed organisms is insufficient for their being moral subjects. The reason is that not all goals are sufficiently important to give rise to human moral duties. She illustrates this with the example of bacteria. While it may be the (unconscious) goal of each bacterium to survive and multiply, it is not evident that we ought to be morally concerned about the goals of individual bacteria. Bacteria do not experience pain, frustration, or grief if their goals are thwarted. They do not care whether or not they survive and multiply. But if they do not care about their own goals, then why should we care about those goals (Warren, 1997)?

Taking the view that beings must be capable of some sort of awareness, sentience, or consciousness if we are to have moral duties towards them, is not tantamount to holding some arbitrary belief. An important reason is that we need such requirements to achieve a tenable conception of our moral duties. After all, our moral thinking about how we as human beings ought to act has two unavoidable limitations. First, it can only be based on what we know and what we care about or ought to care about, and second, our moral concern cannot be extended equally to all of the living beings that exist in the world. But if we in our thinking about moral duties need to be guided by criteria such as awareness, sentience, or consciousness, then we must admit that life is not the only viable criterion of moral status.

To conclude, we have seen that Taylor's theory creates two difficulties. First, its radical egalitarianism has impractical consequences. It has, for instance, too little power to draw distinctions necessary for resolving conflicts between duties to different kinds of living beings. Second, its thesis that life is the only valid criterion of a being's moral status conflicts with a considered conviction in moral thinking. The content of this conviction is that having moral duties towards a being requires more than that the being is a living organism. It requires, for instance, that the being is capable of some sort of awareness, sentience, or consciousness. Because of these difficulties with Taylor's theory we cannot agree to Taylor's defence of respect for life.

2.3.3 Modifications of Taylor's claims?

We may hold that respect for life is a worthy ideal, provided that it is not conjoined with the two claims we have criticized. For this reason we may ask whether the difficulties we discussed above could be reduced or even avoided. We may consider making some modifications to Taylor's claims, for instance the following. We could try to alter the view that life is the only criterion of a being's moral status; we could claim that for there to be moral duties towards a being it must also have some sort of awareness. In consequence we could alter the claim of radical biological egalitarianism. We could inside the set of living beings draw a distinction between on the one hand those beings which 'pursue their own good' in the sense that realizing the good matters to them because they are capable of being aware that the thwarting of it is a state to be avoided, and on the other hand those beings that cannot in this sense pursue their good. Human beings, then, would have moral duties towards human beings and animals but not towards plants.

Making these modifications to Taylor's claims would lead to a reduction of the difficulties discussed above. But it would be open to two objections. The first objection is that in actual fact the modifications amount to radical changes in Taylor's theory. To introduce an additional criterion of moral status, for instance awareness, and to state that we have moral duties only towards living beings capable of awareness, is incompatible both with Taylor's concept of respect for nature and with his biocentric outlook, especially his principle of species-impartiality which says that no bias in favour of some species over others is acceptable. The second objection to the modifications is that they help insufficiently to resolve conflicts of moral duties towards different kinds of beings that are capable of awareness. The modifications transform Taylor's radical biological egalitarianism into an egalitarianism with regard to all beings capable of awareness. But they do not enable us to draw distinctions inside this class of beings. However, we need to draw such distinctions in order to be able to decide how we ought to act when moral duties towards different beings conflict. If we have to make such decisions, we often must take account of considerable differences between beings capable of awareness. There are not only differences as to their own good but also differences in their pursuing this good. For example, some animals seem only to have a rudimentary awareness that realizing their good matters to them, while others can even be said to have intentions. So, thwarting the realization of the latter animals' good may be worse than thwarting that of the former. In sum, the result of the two objections referred to is that the attempt at modifying Taylor's claims fails.

2.4 McMahan's view of killing animals and the badness of death

We have seen that Taylor's theory creates difficulties we cannot accept. This gives rise to the question of whether one can avoid these difficulties and yet defend the position that killing animals is morally wrong because their life has value. For an affirmative answer to this question we can turn to an outstanding study of the ethics of killing in which Jeff McMahan also deals with the wrongness of killing animals (McMahan, 2002). This account says, in short, that killing animals is fundamentally wrong, because it deprives them of goods they would have enjoyed if not killed, or, in other words, it is fundamentally wrong because of the badness of their death.

Before going into McMahan's account in more detail, we should notice that this account in one respect differs fundamentally from Taylor's approach: McMahan's concept of the value of life differs from Taylor's. According to Taylor, the phrase that 'the life of animals has value' should be taken to mean: Animals as living beings have inherent worth. Taken in this sense, the value of a life is the value or worth of the being whose life it is. It is determined by the nature of the subject of the life, by the properties that make the subject what it is: a living being. The lives of all living beings are of equal value or worth. According to McMahan's account, the value of a life is to be distinguished from the worth of the being whose life it is. The value of a life is determined by the character of the contents of the life. It is the value that the life has for the being who lives it. It is equivalent to the extent to which the life is worth living. This concept does not imply that the lives of all animals have equal value. It does justice to the fact that some animals experience much less good than others.

2.4.1 Two basic ideas

Let us now look more closely at McMahan's statements about the wrongness of killing animals. A very short phrasing of this account has been mentioned above: Killing animals is fundamentally wrong, because it deprives them of goods they would have enjoyed if not killed. This formulation already suggests two basic ideas. First, McMahan speaks of the wrongness of killing animals, because he regards animals as beings which are capable of consciousness and have a stake in how their futures go. Second, his judgement on killing is based on his view concerning the badness of death for the victims. Inspired by Nagel (1993) and others, McMahan takes the view that the badness of death for the individual who has died can be accounted for by appeal to what this individual has been deprived of. To put it another way, he takes the view that the evaluation of the badness of death for the individual who has died is dependent on how the life of that individual would have gone had it not died.

If one takes this view, then one gets the problem of how one should evaluate the life the individual would have had if it had not died. In order to clarify this complex problem, McMahan presents two rival evaluations of the badness of death. He calls them the 'Life Comparative Account' and the 'Time-Relative Interest Account of the badness of death' respectively. Let us now turn to these two accounts.

2.4.2 Two accounts of the badness of death

The Life Comparative Account evaluates the badness of death for the individual who has died by considering the goods that would have been realized in its life had it not died. The badness of death is measured in terms of its effect on the overall value of the life as a whole. A particular death is the worse, the greater the difference between the total value the life as a whole would have if the death were to occur and the total value the life would have if the death were not to occur.

The Time-Relative Interest Account of the badness of death evaluates death in terms of the effect it has on the individual's time-relative interests rather than on the value of the individual's life as a whole. That the individual has a time-relative interest in continuing to live means that it matters for the individual's sake now or from its present point of view, that he should continue to live. How strong this interest is depends on two factors: first, what amount of good the individual's life would contain if this individual were not to die, and second, how strong the psychological connections would be between the individual now and the individual in the future, assuming it were to live. The weaker the psychological connections between the individual at the time of death and the individual at the time the good in question would have been realized in its life, the less important that good is for evaluating the extent to which the individual's death was bad for it. The Time-Relative Interest Account holds that the badness of death for the individual is proportional to the strength of the individual's time-relative interest in continuing to live.

2.4.3 Two accounts of the wrongness of killing

According to McMahan, the difference between these two accounts is of practical importance. For he regards the rival evaluations of the badness of death as foundational for two different views of the wrongness of killing. He calls these views the 'Harm-Based Account' and the 'Time-Relative Interest Account of the wrongness of killing' respectively. Let us take them in this order.

McMahan states that the Life Comparative Account of the badness of death is closely connected with widespread reflections about the morality of killing. The initial thought in such reflections is that killing is wrong because of the dreadful effect it has on the victims: it deprives them of the good life they would otherwise have had. McMahan regards this initial thought as the foundation of what he calls the 'Harm-Based Account of the wrongness of killing'. This account holds that acts of killing are normally wrong principally because of the harm they inflict on the victims, and that the degree to which an act of killing is wrong varies with the degree of harm it causes to the victim. In this account, the view that killing is wrong because it involves the infliction of harm on the victim is combined with the assumption of correlative variation, that is to say, the assumption that the greater the harm an act of killing inflicts on the victim, the more seriously wrong it is. So, the account states that killing persons is usually gravely wrong, for death is typically among the worst harms a person can suffer. It also states that the killing of animals is in general less seriously wrong than the killing of persons, and that the killing of animals of a certain type is generally more seriously wrong than the killing of animals of other types. Because animals vary considerably in their capacities for well-being, some may be harmed to a greater extent by death than others. McMahan illustrates this with the example of dogs and frogs. Because a dog's life is normally richer in pleasure, social relations, and so on, than a frog's, dogs generally suffer a greater harm in dying, and it is normally more wrong to kill a dog than it is to kill a frog.

On the basis of the Time-Relative Interest Account of the badness of death McMahan formulates a corresponding Time-Relative Interest Account of the wrongness of killing. This account holds that what is fundamentally wrong about killing is that it frustrates the victim's time-relative interests. Like the Harm-Based Account, it explains what is fundamentally wrong about killing in terms of the effects on the victim. But it insists that it is the psychological connections which may hold to varying degrees, that matter. It also incorporates its own assumption of correlative variation, namely, that the degree to which an act of killing is wrong varies with the strength of the victim's time-relative interest in continuing to live. The Time-Relative Interest Account of the wrongness of killing, then, holds that it is a much more serious wrong to kill a person than the killing of an animal. The reason for this is that an animal's time-relative interest in continuing to live is weaker than a person's. It is weaker because of two facts. First, the amount of good that an animal loses through death is much less than that which a person loses. For example, there are good experiences and actions in a person's life which require complex reasoning abilities that, to the best of our knowledge, animals do not possess. Second, the psychological connections that would bind an animal to itself in the future are weaker than those in the life of a person. For example, a person can have long-range desires or longstanding projects which are absent in the life of an animal. However, it is important to notice that these comparative judgements do not invalidate the claim that killing an animal is morally wrong because it frustrates its time-relative interest in continuing to live. The animal too would lose future good through death, even though it would lose less than a person would lose and even though its psychological connections to itself in the future would be weaker than a person's. Moreover, the Time-Relative Interest Account takes into consideration that in both respects there are important differences between animals. Therefore, it also holds that the killing of a more complex animal is more seriously wrong than killing a simpler one, because a more complex animal's time-relative interest in continuing to live is stronger than a simpler animal's.

2.4.4 Which account should we endorse?

We have now two accounts of the badness of death and two accounts of the wrongness of killing. These accounts express McMahan's endeavour to clarify the problems of how to evaluate the badness of death and the wrongness of killing. But they do not yet settle these problems. The question arises which of the rival accounts we should endorse. McMahan answers this question by pleading for the Time-Relative Interest Accounts of the badness of death and the wrongness of killing. A clear instance of this is to be found in his discussion of the wrongness of killing. There he takes the view that the Time-Relative Interest Account is superior to the Harm-Based Account. His reasoning can be presented in two steps. The first step is that the Harm-Based Account faces an objection that undermines it. The Harm-Based Account states that the harm involved in death is equivalent to the extent to which the death makes the individual's life as a whole worse than it otherwise would have been. So, the Harm-Based Account presupposes the Life Comparative Account of the badness of death. But the Life Comparative Account has some profoundly counterintuitive implications. For instance, it appears to yield the wholly implausible conclusion that the death of an infant is worse for the infant than the death of a middle-aged person for that person because the infant has been deprived of more of the goods that life has to offer. The Time-Relative Interest Account avoids this conclusion because it insists that the psychological connections between the individual now and the individual in the future are of great importance. Though the infant has been deprived of more goods, its psychological connections with the future stages of her life are quite weak. So, at the time of death the infant does not have particularly strong reasons to care about many of the goods she has been deprived of by death. But the same cannot be said of the middle-aged person. The second step in McMahan's reasoning is a direct moral argument in favour of the Time-Relative Interest Account. By stressing the importance of the psychological connections between the individual now and in the future, the Time-Relative Interest Account directs our concern to what the individual now has a stake in, that is, what it has most reason to care about, and this concern accords with what morality requires, namely, to be concerned for an individual for his own sake. So, according to McMahan, we should follow the Time-Relative Interest Account.

However, there is a limit to the Time-Relative Interest Account. Taken as an account of the badness of death, McMahan endorses it. But taken as an account of the wrongness of killing, he thinks it valid only with regard to killing animals but not with regard to killing persons. For the wrongness of killing a person cannot be adequately accounted for solely by appeal to her frustrated time-relative interest. People differ widely in strength of their time-relative interest. In those people whose futures promise a great deal of good, the time-relative interest is strong; in others, it is comparatively weak. The Time-Relative Interest Account of the wrongness of killing would hold that killing people of the latter sort is substantially less wrong morally than killing people of the former sort. But this implication profoundly offends the moral conviction that under normal circumstances all killings of persons are equally wrong. This conviction is supported by the idea that all persons have autonomous wills and therefore have a moral standing that demands equal respect, independent of the goods their lives may contain. To do justice to this conviction, McMahan introduces what he calls the 'Two-Tiered Account of the wrongness of killing'. This account distinguishes between persons and animals. Killing a person is normally wrong because it constitutes a failure to respect the victim's status as a person. The seriousness of wrongfully killing a person does not vary with the strength of her time-relative interest in continuing to live. But the wrongness of killing an animal who is not a person has to do with the frustration of the animal's time-relative interest in continuing to live.[1]

2.4.5 Conclusions

We are now in a position to say what we, according to McMahan, should understand by the thesis that killing animals is morally wrong because their life has value. He puts forward that we should take it to mean that killing animals is fundamentally wrong because it frustrates the animals' time-relative interest in continuing to live. This statement contains three claims. The first is that killing an animal is wrong, because it deprives the animal of future good which it would have enjoyed if not killed. This loss is bad for the animal, because the animal is capable of consciousness and has a stake in how its future goes. Prospective future good matters to it. So, the animal is harmed or wronged by being killed. The second claim is that killing some animals is more seriously wrong than killing others, because the former have a stronger time-relative interest in continuing to live. They lose a greater amount of good through death and their psychological connections to themselves in the future are stronger. So, the killing of a more complex animal is more seriously wrong than killing a simpler one. McMahan's third claim is that it is sometimes defensible to go against the animal's time-relative interest in continuing to live. But if one is to be justified in killing an animal, one must show that killing it is necessary in order to prevent even greater harm or to produce benefits that outweigh the harm inflicted. McMahan (2003) considers whether two widespread practices can pass this test. The first is that of rearing and killing animals for food. He brings forward that this practice is unjustified because the benefits it offers us do not outweigh the harms it inflicts on animals. He argues as follows. The difference between the pleasure from eating meat and the pleasure we could get from a wholly vegetarian diet is negligible. The argument that abolishing the meat industry would be prohibitive can be advanced on behalf of any large-scale

---

[1] McMahan's extensive investigation into the permissibility of killing human beings and criticisms of it, notably by Kumar (2006), will be left out of consideration here, because our topic is killing animals.

practice, however iniquitous. A different practice could be justified if the animals were treated differently, but this is not a realistic policy outside small, rural settings. The second widespread practice McMahan considers includes xenotransplantation (the transplantation of organs taken from animals into the bodies of human beings) and various forms of experimentation. Though this practice also involves harming and killing animals, he finds it more likely to be defensible because the benefits it offers are very significantly greater than those we derive from eating meat.

We may conclude that McMahan's statement avoids the difficulties caused by Taylor's theory mainly because McMahan's understanding of the value of life, unlike Taylor's, does not hinge on the morally obligating and egalitarian concept of the inherent worth of living beings but focuses on the value that the life has for the being who lives it.

References

Kumar, R., 2006. Permissible killing and the irrelevance of being human. Journal of Ethics 12: 57-80.

McMahan, J., 2002. The ethics of killing: problems on the margins of life. Oxford University Press, New York, NY, USA, 556 pp.

McMahan, J., 2003. Animals. In: Frey, R.G. and Wellman, D.H. (eds.) A companion to applied ethics. Blackwell, Oxford, UK, pp. 525-536.

Nagel, T., 1993. Death. In: Fischer, J.M. (ed.) The metaphysics of death. Stanford University Press, Stanford, CA, USA, pp. 59-69.

Rollin, B.E., 1992. Animal rights and human morality. Prometheus Books, Amherst, MA, USA.

Rollin, B.E., 1995. The Frankenstein syndrome: ethical and social issues in the genetic engineering of animals. Cambridge University Press, Cambridge, UK, 256 pp.

Singer, P., 1993. Practical ethics, 2nd edition. Cambridge University Press, Cambridge, UK, 406 pp.

Taylor, P.W., 1986. Respect for nature: a theory of environmental ethics. Princeton University Press, Princeton, NJ, USA, 329 pp.

Warren, M.A., 1997. Moral status: obligations to persons and other living things. Oxford University Press, Oxford, UK.

3. Killing as a welfare issue

R.P. Haynes[1]
Emeritus Professor of Philosophy, University of Florida, 6802 SW 13th St., Gainesille, FL 32611, USA

Abstract

In this paper, I will argue that, under normal circumstances, killing an animal robs it of something crucially important to it – the ability to enjoy the good things of life. From this perspective, as Sapontzis (1987) convincingly argues, life has important instrumental value to the animal (or human) that possesses it. I will briefly identify arguments against Sapontzis's position – that killing cannot be a harm because the victim no longer exists and so cannot feel or be harmed. I will also argue that killing deprives the victim of its welfare. To argue for this position, I will adopt Sumner's account of human welfare, and then apply it to animals. As an aside, I will comment on the questions, under what circumstances are we morally justified in killing an animal and what obligations do we have, if any, toward animals not under our care.

Keywords: animal welfare, death, justice, caregivers

3.1 Does killing harm the victim?

Intuitively, we should say that it does: killing harms the victim. Independently of Sapontzis's claim that life has instrumental value and so killing deprives the victim of this important instrument that enables a person (animal) to enjoy the good things of life. Normally, we would consider killing someone as the greatest harm that could be rendered to them. I can think of two arguments against the claim that killing an animal harms that animal. One is that only humans can consciously value life, so depriving a non-human animal of life does not deprive it of something that it values. The other argument is that by 'harming someone or thing' we mean that the harmer has made the victim or object lose something that it valued or, in some sense, made it worse off. But if we believe, as I do, that a dead person or animal no longer exists as a person or animal, then it make no sense to say that it is worse off than it was before the harm. This makes sense to me and is a good reason for not fearing death. Perhaps humans fear death, independently of being concerned how my death would affect those who value my existence. The reason is that if what makes my life worthwhile is the projects that I am in the process of creating, or my friendship that I enjoy because I believe that my friends value my existence, then it is difficult to continue valuing these project if we know that we will not be here to appreciate them.

---

[1] Professor Haynes passed away in 2014

Franck L.B. Meijboom and Elsbeth N. Stassen (ed.) **The end of animal life: a start for ethical debate**
DOI 10.3920/978-90-8686-808-7_3, © Wageningen Academic Publishers 2016

As far as the argument that only humans can value life, this seems counter-intuitive if we consider that animals also devote time to creating things – off-springs, for one. So if we accept the idea that death is a harm for humans, then I can see no reason for not extending this to non-humans.

Next, let us consider whether death reduces the welfare of its victim. First, however, we must be clear about what we mean by 'welfare'.

3.2 What does 'welfare' mean?

Although the word 'welfare' has come to refer to a system of caring for disabled citizens, in the context of discussions about animal welfare, it clearly refers to what, in the case of humans, we might call 'well-being' or 'happiness'. To be clear about how to conceptualize welfare in the context of far-reaching discussions about animal welfare, I suggest we start with a concern for how to think about human welfare. When do we rightfully consider a person to be well-off or happy? In what does their welfare consist? In his book, Sumner has addressed this question in the context of those who argue that it consists of having one's preferences (or desires) satisfied, regardless of what these preferences or desires are, or that it consists of an objective list of goods and the possession of the items on this list make a person well-off whether they realize it or not. Sumner's position is that a person is well-off to the extent that they are satisfied with how their life is going, providing that this satisfaction is justified (Sumner, 1995). This qualification is introduced to preclude calling a person well-off who is satisfied with their life because they have been socialized to think that they are worthy of nothing more than what they have. So a person is well-off to the extent that they are justifiably satisfied with their life. Although Sumner is reluctant to apply this to non-human animals, I see no reason why we cannot use it with the proviso that humans make the judgment whether an animal that seems to be satisfied with its life is aware of all of the reasonably expected elements that could be added to its life. If an animal is experiencing a life that is as good reasonably as could be expected for an animal of that type, then we are entitled to judge that the animal is justified in being satisfied with its life.

3.3 Preference satisfaction as the criterion for a good life

Some 'animal welfarists'[2] seem so concerned with tests to determine whether particular animals are well-off that they seem to have settled on the use of 'preference tests', seemingly relying on the preference satisfaction criterion that Sumner (1995) rejects. Sumner discusses problems with a preference theory of welfare by classifying it as a theory of desire satisfaction.[3] The principle reason for settling on the preference

---

[2] Largely animal scientists concerned with addressing the animal welfare issues, e.g. Fraser and Matthews, 1997.

[3] For scepticism about the limited value of preference tests, see Fraser and Matthews, 1997.

criterion, I believe, is that it preserves a role for animals welfare scientists to measure how well off animals kept in captivity to produce food can be said to be well off even in the limited environment in which they are raised. One of the major problems with preference tests, in the context of animals kept for food purposes is that they are usually conceived of in terms of which of two situations the animal prefers, such as straw or wood flooring in the housing. An animal may choose the least undesirable alternative, but if given more options would choose neither of the two offered in the previous test. In order to be a legitimate test, the chooser would have to be aware of all of the reasonably possible alternatives, and choose the ones preferred under those conditions. There is still the possibility that given the animal's past, they may think one or more of these is best, when if socialized differently would make other choices. A qualified judge of an animal's welfare would need to know a lot about that particular species and also about that individual animal. There are few conditions in existence that would provide such information, although some have suggested an ideal 'pig park' as a way of finding out about what captive pigs would prefer (e.g. Wood-Gush and Stolba, 1981). However, one may question the relevance of this proposal, because even the life of pigs in the wild provides an inadequate measure of what a truly concerned caregiver might add to that life.

Another, more theoretical objection to a preference theory of welfare is that it does not take account of the role of socialization in determining one's preferences or even desires, and we might have been socialized to prefer or even desire things that are not really good for us. And it is one thing to say that having one's desire satisfied is a major criterion of the good life, and another to focus on preferences because if asked to choose between two situations in the world, I might prefer one to the other, but I do not actively desire either one.

3.4 What do we owe animals under our care?

If we are an animal's caregiver, then that animal is our ward. It is common to refer to farm animal keepers as caretakers. But caretakers usually take care of property, whereas caregivers have a relationship similar to that between a parent and child or a pet (Kheel, 2004) If we choose to be responsible for the sort of life that animal is going to live, then there is little difference between that choice and the choice to have children. Both come with certain responsibilities. Thus we owe animals what we owe other wards, such as the children we raise, in other words, the best life we are capable of providing, otherwise we have no business assuming the wardship of that animal. Someone might argue that our duties to our children are based on the social good a well raised child might serve, and the social evil it might do if poorly raised. I suppose this is a serious objection, but it does not really apply to our pets, that we love and want the best life for. But from the point of view of a virtue ethics, caregivers who would be indifferent to the wellbeing of their wards when they are raised for food rather than as pets, we would judge such a caregiver as seriously lacking in sympathy for another living thing and how it might suffer.

What does this imply about using that animal for our own profit? While harvesting the eggs of our chickens may not interfere with the chicken's welfare, killing it for food clearly does. Killing that animal interferes with its welfare, unless it is suffering from an incurable ailment, since killing is a harm, as I have already indicated in the first section. Some have argued that under certain conditions, it is still acceptable to kill animals under our care. Some have employed for this purpose the 'contractual model' of our relationship to animals farmed for food. Let us next consider their arguments.

3.5 The contractual model

The contractual model makes the claim that suitably reformed farming practices that lead to the eventual slaughter of animals for food consumption are a fair deal for animals involved. There are several versions of this model. One version is that there is a tacit contract between domestic animals and their owners or custodians. Another version is that since we bred them, they owe their lives to their breeders. Somewhat different versions are that a fairly short agreeable life is better than none at all, or that domestic life is better than life in the wild. Finally, there is the argument that in exchange for being protected from wild predators, the protectors need to function themselves as predators in order to manage the size of the population. Then there is the 'neoteny' thesis to support the claim that some sort of choice was made by the ancestors of domestic animal to become domestic and dependent on the care of their caregivers (Budiansky, 1992; Coppinger and Smith, 1983). For example, some animals, such as dogs or cats or even cattle, have actually taken advantage of situations where they can profit from having a relationship with humans, and then this trait was bred into their offspring over the centuries. In this sense, they remain perpetually like children seeking caregivers. This would imply that this type of animal chooses humans as their caregivers, thus relieving humans of the obligation that they would otherwise have if it were us who chose the relationship and sponsored their breeding.[4]

There are a number of ways that these arguments can be addressed. For example, contracts are usually invalid when some of the parties are unaware of the conditions of the agreement, or, we have bred our children but do not consider that it is alright to kill them. And as far as the argument that domestic life is better than life in the wild, Kheel (2004) citing Lackner (1984) points out that only about 5% of wild animals are killed by other animals. Furthermore, it is difficult to make sense of the choice, would

---

[4] For different versions of the contract arguments, see for example, Appleby (1999), Larrère and Larrère (2000), Lund *et al.* (2004).

you rather live a short life than none at all.[5] Which brings us to the question of what we owe animals not under our care? Clearly, killing them harms them, but is that a moral wrong?

3.6 What do we owe other animals?

Now that I have given my arguments for our obligations to animals under our care, I think we should address the problem of how we should treat animals not under our care and to whom we do not have a caregiver relationship. This is a difficult question to answer. Given that killing them harms them, is it wrong to kill them? And are we morally obligated to protect them from other harms? The only way to answer this question that makes sense to me is from a virtue ethics perspective. Of course, we could argue that animal life enriches our environment, and thus our own lives (Nussbaum, 2004; O'Neill *et al.*, 2008), but are we morally obligated to do that? And some animals are more a nuisance than an enrichment. My case would be that a human that is sensitive to the suffering of other animals is a better person, and being that better person enriches their own life, just as being morally good does.[6] If we think that it is advantageous to be that kind of person because it makes us a better person, then we have solved part of the problem. But we still have to face up to the problem about how to manage the conflicts between living things, such as the conflict between predators and their prey, or even the conflict between us and animals that feed on our crops or ornamentals. What is the right or just thing to do in regard to protecting or not protecting their welfare? To answer these questions, I propose a general theory of justice for nonhuman animals (see also Haynes, 2008).

3.7 A general theory of justice for nonhuman animals

The main purpose of the practice of administering justice is to protect the weak and vulnerable and protect them from harm. This is a position that Sapontzis (1987) develops and that Nussbaum (2004) also argues for. Nussbaum's version is that all animals have an equal right to lead a flourishing life. This claim seems intuitively sound to me. The principle implies that humans ought not to interfere in an animal's ability to flourish. But what if there are other impediments to an animal's flourishing? Are humans obligated also to try to remove these impediments where they can? One such impediment to animals living in the wild is predation. Are we obligated to prevent predation, even if there is a conflict between the prey's flourishing and the predator's flourishing? And what are we obligated to do when the flourishing of particular animals is destructive of wild-life habitat, such as wild pigs that root up large portion of wild lands or of an environmentally healthy environment because their predators no longer

---

[5] For a more detailed account of the contract and neoteny arguments, who has held them, and their problems, see Haynes (2008).

[6] Cf. the virtue ethics literature

Kheel, M., 2004. Vegetarianism and ecofeminism: toppling patriarchy with a fork. In: Spaontzis, S.F. (ed.) Food for thought. The debate over eating meat. Prometheus, Amherst, NY, USA, pp. 327-341.

Lackner, S., 1984. Peaceable nature: an optimistic view of life on earth. Harper and Row, San Francisco, CA, USA.

Larrère, C. and Larrère, R., 2000. Animal rearing as a contract? Journal of Agricultural and Environmental Ethics 12: 51-58.

Lund, V., Anthony, R.E. and Röcklinsberg, H., 2004. The ethical contract as a tool in organic animal husbandry. Journal of Agricultural and Environmental Ethics 17: 23-49.

Manning, R., 1996. Caring for animals. In: Donovan, J. and Adams, C.J. (eds.) Beyond animal rights. A feminist caring ethic for the treatment of animals. Continuum Books, New York, NY, USA, pp. 103-125.

Midgley, M., 1983. Animals and why they matter. University of Georgia Press, Athens, GA, USA.

Nussbaum, M.C., 2004. Beyond 'compassion and humanity': justice for nonhuman animals. In: Sunstein, C.R. and Nussbaum, M.C. (eds.) Animal rights: current debates and new directions. Oxford University Press, New York, NY, USA, pp. 299-320.

O'Neill, J., Holland, A. and Light, A., 2008. Environmental values. Routledge, New York, NY, USA.

Sapontzis, S.F., 1987. Morals, reason, and animals. Temple University Press, Philadelphia, PA, USA.

Sumner, L.W., 1995. Welfare, happiness, and ethics. Clarendon Press, Oxford, UK.

Wood-Gush, D.G.M. and Stolba, A., 1981. Behavior of pigs and the design of a new housing system. Applied Animal Ethology 8: 583-585.

4. Death, *telos* and euthanasia

B.E. Rollin

Department of Philosophy, Colorado State University, Fort Collins, CO 80523, USA;
bernard.rollin@colostate.edu

Abstract

Since Bentham, animal ethics has to a large extent been based in Utilitarianism, maximizing pleasure and avoiding pain. But the ability to feel pain, while sufficient for a being to obtain moral status, is not a necessary condition. What is necessary for moral status is that what happens to or is done to a being *matters to that being*, in either a negative or a positive way. In our world, however, most of the 'mattering' necessary to survival is negative – injuries and unfulfilled needs ramify in pain. But physical pain is by no means the only morally relevant mattering – fear, anxiety, loneliness, grief, certainly do not equate to varieties of physical pain, but are surely forms of 'mattering'. Indeed, an adequate morality towards animals would include a full range of possible matterings unique to each kind of animal. In my account of animal ethics, I have argued that the basis of our obligations to animals under our aegis is the animal's nature, what I call *telos* following Aristotle. This is the unique set of traits and powers that make the animal what it is – the 'pigness' of the pig, the 'dogness' of the dog. Some *telos* violation matters more than pain. Happiness may be understood as satisfaction of needs flowing from animal *telos*. The moral import of death is discussed in relation to *telos*, pain, and euthanasia.

Keywords: positive mattering, negative mattering, animal nature

4.1 Pain and *telos* as the bases for animal ethics

Thanks to a series of British 'commonsensical' thinkers, animal ethics, at least in the English-speaking world, has focused overwhelmingly on pleasure and pain. These thinkers generally worked in the Utilitarian tradition and included Bentham, Mill, Sidgewick, Salt, and Singer, as well as related philosophers like Hume (1779). Being empiricists, it seemed to them that the desire on the part of all organisms to seek pleasure and avoid pain was an observable and obvious basis for ethics that lent itself well to quantification. This focus enabled British moral thought to escape from continental European, rationalistic scepticism about more complex animal thought to exclude animals from the realm of moral concern, as articulated by Descartes, Spinoza, and Kant. Animal ethics could thus be grounded in common sense awareness that animals experienced pleasure and pain even as we did; that pleasure and pain *matter* to animals.

*Franck L.B. Meijboom and Elsbeth N. Stassen (ed.) **The end of animal life: a start for ethical debate***
DOI 10.3920/978-90-8686-808-7_4, © Wageningen Academic Publishers 2016

As Bentham (1789) put this point:

> Other animals, which, on account of their interests having been neglected by the insensibility of the ancient jurists, stand degraded into the class of things. ... The day has been, I grieve it to say in many places it is not yet past, in which the greater part of the species, under the denomination of slaves, have been treated ... upon the same footing as ... animals are still. The day may come, when the rest of the animal creation may acquire those rights which never could have been withheld from them but by the hand of tyranny. The French have already discovered that the blackness of skin is no reason why a human being should be abandoned without redress to the caprice of a tormentor. It may come one day to be recognized that the number of legs, the villosity of the skin, or the termination of the os sacrum, are reasons equally insufficient for abandoning a sensitive being to the same fate. What else is it that should trace the insuperable line? Is it the faculty of reason, or perhaps, the faculty for discourse? ... The question is not, Can they reason? nor, Can they talk? but, Can they suffer? Why should the law refuse its protection to any sensitive being? ... The time will come when humanity will extend its mantle over everything which breathes ...

Bentham could not have imagined that Scientific Ideology, a set of unexamined dogmatic beliefs about the nature of science, would eventually vanquish empiricist common sense and instead create agnosticism regarding animal consciousness and animal pain, even among those scientists who used animal models to study human pain! For now, it is necessary to ask a neglected question: does acknowledging pain in animals provide a sound and solid basis for including animals within the moral arena?

Before answering this question, we should acknowledge that, in every human society, actions that affect other people are limited, circumscribed, and constrained by some moral principles. (Indeed, according to some moral theorists, all such moral principles are at root the same.) And that, in turn, results from the fact that many human actions have effects on others, both helpful and harmful. When such actions are allowed to proceed at random, it becomes impossible to assure a peaceful society consisting of people working together harmoniously, as the strong will always attempt to impose their will on the weak. The creation of morality, then, is essential for peaceful and creative coexistence. Because we are all vulnerable, morality exists to provide protection from one another, as well as to encourage benevolence among us, which might never emerge in the Hobbesian situation of a 'war of each against all'. The most obvious weakness we all share is vulnerability to any pain or harm that others may inflict upon us.

It is for this reason, of course, that utilitarian philosophers based morality foursquare upon the social goal of minimizing pain and maximizing pleasure across society. They further believed that all moral prescriptions could be reduced to that goal. And, since animals could be hurt or benefited by human actions, it was appropriate in their view to include animals within the scope of moral concern.

But the ability to feel pain is not a necessary condition for moral considerability. For example, a person or animal unable to feel pain warning of burns or infection resulting in loss of a limb would still be morally considerable, and we would be blameworthy if we did not help such a person or animal preserve their limb, for example, since being able to walk or run or have two arms very much *matters* to the person. The mattering in question is not of course a question of mere pain. Such a person surely suffers, but all suffering does not equate to pain, unless the word 'suffering' is used simply as a synonym for pain, which is very implausible. If we lose the ability to see or hear or articulate (as for example following a stroke) we suffer, yet there may be no pain involved though there is surely *harm*. The whole point of the Utilitarian emphasis on pleasure and pain is to create a continuous spectrum allowing us to quantitate harm in terms of degree, yet as Mill unwittingly showed, as soon as one talks of different *kinds* of pain (for example emotional versus physical pain), something besides pure pain assumes moral relevance.

Or, to take a more forceful example, Hume (1779) pointed out that organisms could have possibly evolved so as to be motivated to flee danger or injury or to eat or drink not by avoidance of pain, but by 'pangs of pleasure' that increase as one fills the relevant need or escapes the harm. In such a world, 'mattering' would be positive, not negative, but would still be based in sentience and awareness.

In our world, however, most of the 'mattering' necessary to survival is negative – injuries and unfulfilled needs ramify in pain. But physical pain is by no means the only morally relevant mattering – fear, anxiety, loneliness, grief, certainly do not equate to varieties of physical pain, but are surely forms of 'mattering'. This was recognized in US laboratory animal legislation in its demand for control of 'distress', a catch-all term for modalities like the above, not just 'pain'.

Indeed, an adequate morality towards animals would include a full range of possible matterings unique to each kind of animal. In my account of animal ethics, I have argued that the basis of our obligations to animals under our aegis is the animal's nature, what I call *telos* following Aristotle (Rollin, 2006). This is the unique set of traits and powers that make the animal what it is – the 'pigness' of the pig, the 'dogness' of the dog. This is well recognized in common sense, exemplified in the song affirming that 'fish gotta swim, birds gotta fly'. If we raised pigs, for example, under totally natural conditions, satisfying all aspects of pig nature, from nest-building to rooting, we could say we understand 'happiness' relative to that animal. Happiness is satisfaction of *telos* needs. When we fail to meet needs flowing from the *telos*, we harm the animal. While we do not have a word for the mattering implicit in failing to allow a pig to forage, or build its nest, as we keep them in modern confinement agriculture we can plainly see that each of these failures to meet what the animal is by nature is going to create a harm we are guilty of committing. The word 'pain' simply does not capture the myriad ways different treatments affect animals.

Sometimes not meeting other aspects of animal nature matters more to the animal than does physical pain. Kilgour reported that cattle show more signs of stress when introduced into a herd of strange animals than when they are prodded with an electric prod (Kilgour, 1978). Chickens will go through electro-shocking grids to get access to the outdoors. Ethologists have given us myriad such examples. Whatever a calf and cow feel when they are separated shortly after the calf's birth, and the cow moos for months thereafter, it is not physical pain but something that clearly causes suffering.

An example from coyote behaviour strikingly illustrates how *telos* needs can trump even major physical pain. It has been recounted for years that a coyote, caught in a leg-hold trap, will chew its leg off, enduring terrible pain, rather than submit to immobility (this is also true for other animals, such as raccoons). This is understandable given the coyote's *telos* as a free ranging predator (or, on occasion, prey). It is not plausible to suggest that the animal chews its leg off to avoid death, since it is not possible that a non-linguistic being *has* a concept of death, though it clearly understands the inability to escape. As Heidegger (1996) remarked, to understand death requires that one grasp the 'possibility of the impossibility of one's being,' a notion involving a possible state of affairs, which, as a counterfactual, requires sophisticated syntax that there is no reason to believe animals possess. For similar reasons, one cannot communicate to an animal that 'there are no dragons in the library,' as opposed to 'there are no chickens or cats in the library'.

There is no simple word to express the many ways we can hurt animals besides creating physical pain; the ways are as countless as the multiplicity of *teloi* and the interests that flow from them. So in this essay I will introduce a barbarous neologism to express this concept – 'negative mattering'. 'Negative mattering' means all actions or events that harm animals – from frightening an animal to removing its young unnaturally early, to keeping it so it is unable to move or socialize. Physical pain is perhaps the paradigmatic case of 'negative mattering', but only constitutes a small part of what the concept covers. 'Positive mattering' would of course encompass all states that are positive for the animal – freedom of movement, pleasure, a sense of security, and so on.

If our analysis is correct, it is morally obligatory to expand the scope of veterinary medicine and/or animal welfare science to study all of the ways things can matter negatively to animals as well as positively matter, as society grows ever more concerned about animal treatment. In addition, it is necessary to attempt to understand which forms of negative mattering are most problematic from an animal's perspective. In cattle, we have seen, surprisingly, that being exposed to a new herd with no preparation has more of a negative effect on cattle experience than does electric shock. We have seen that immobilization is more aversive to a coyote than is significant pain. Obviously the challenge is to study these behaviours without hurting the animal subjects. An excellent example of this issue arises from research into 'learned helplessness'. Research in this area is so barbaric it has been banned in the UK for some time. Absurdly alleged to be a model for human depression, learned helplessness is achieved by subjecting an animal in a cage to inescapable electric shock regardless of what it does. Eventually the animal assumes a foetal position and does nothing. This horrible state rarely if ever arises in nature, and once again, there is no name for the feeling engendered. Ironically, now that we know about it, we can assume that it occurs in animals housed in severely a restrictive environment such as sow stalls which replicate helplessness.

Also necessary to study is the way in which the quality of negative experiences changes with cognitive states of the animal. As both Weiss (1972) and Mason (1971) have shown, the animals' mental and cognitive states regarding negative experience modulate the degree to which the animal experiences the event in question as negative. Mason's work demonstrated that elevation of ambient temperature of mice's environmental surroundings to well above the comfort zone varies in the degree to which the animal is disturbed in accordance with whether the elevation is gradual (and thus is cognitively processed by the animals as predictably rising) or sudden, where the animal has not had the ability and time to adjust its expectations to the advent of unpleasant ambient temperature or 'heat stress'.

Similarly, Weiss showed that monkeys who are taught to anticipate and predict an electric shock, for example by a bell sounding prior to the shock, have far less of a negative reaction to the shock than do those who do not know when the shock is coming. These studies have profound implications for the non-pharmacological control of pain. It has long been known that laboratory animals subjected to an invasive procedure that is followed by a reward have less of a negative reaction to the procedure than those who are simply restrained. In some cases, researchers train animals to a procedure. In one instance, my friend was drawing blood from dogs daily for a vaccine study. She would enter the facility, play with each dog, draw the blood, and give the dog a treat after the draw. On one occasion, one of the dogs set up such a howl as she was leaving that she raced back to see if his paw was caught in the cage door. It turned out she had forgotten to draw blood from that dog, and he had missed his play and his treat, which bothered him more than the blood draw.

It follows from what we have been arguing that invoking *telos* as a basis for animal ethics is far more satisfactory than restricting one's considerations to the categories of pleasure and pain. It is odd to affirm that keeping animals like pigs, built to move or to forage for a mile a day, in a gestation crate at all times causes *pain* to the animal, except in a metaphorical sense. In contrast, it is perfectly clear to ordinary people, and common sense when one says that such a system of husbandry *violates the animals' nature.* It is for this reason that I consider *telos* to capture much better than do pleasure and pain what common sense wishes to embody in animal ethics. Indeed, the concept of the Bill of Rights embodied in US democracy, which serves as the basis for American societal ethics, rests foursquare on the concept of human *telos.* Similarly, animal *telos* most plausibly serves as the basis for articulating the emerging societal ethics for animals, for it is already familiar to us from human ethics, and thus we can, as Plato suggests, *recollect* what we already know.

Let us recapitulate the argument we have advanced thus far. Basing ethical obligations to animals on their ability to feel pain was certainly a conceptual advance over mystical criteria such as possession of a soul, but is open to a number of objections. Most importantly, the ability to feel pain is not a necessary condition for being an object of moral concern, though it is surely a sufficient condition. The ultimate necessary condition for moral status is that what one does to the being in question *matters to it.* There are many ways of mattering, both negative in addition to pain – fear, loneliness, boredom and anxiety, and positive in addition to pleasure – presence of companions, freedom to roam, ability to exercise. Especially important in terms of both positive and negative mattering, are the factors that make up an animal's *telos.* A *telos* represents the unique set of ways that an animal instantiates the functions of living things – nutrition, sensation, reproduction, cognition, acquiring food and water, and so on. If an animal can fully express all elements of its *telos*, there is a good reason to call that animal 'happy'. If animals cannot actualize their *teloi*, to the degree that they cannot constitutes unhappiness or misery. We must also remember that while much of physical pain can be pharmacologically controlled, most of the misery resulting from violation of *telos* cannot.

4.2 Does death per se matter to animals?

It follows from what we have argued that all aspects of negative mattering should enter into the field of pain control. Nonetheless, physical pain has been called 'the worst of evils'. At first blush, this appellation seems wrong, since we are inclined to believe that *death* is the worst of evils. But careful reflection reveals the falsity of such an assertion. In the first place, even among humans, people will readily choose death over prolonged or intense physical pain, and even over emotional pain. This is evident in the world-wide thrust for assisted suicide on the part of those suffering intractable physical pain and/or mental anguish. Indeed, many people choose death over helplessness, or total dependence on others, or loss of dignity through incontinence. While there are many cases such as radical cancer treatment where people undergo prolonged and horrible pain in order to live, there are also very many where death is chosen to forestall suffering.

Since our topic is animal pain, the question arises as to how animals value death as compared with pain. This is an increasingly important conceptual/ethical issue in a world where ever-increasing amounts of highly invasive therapy are being exported from human medicine to animal medicine, resulting in great amounts of suffering. This is particularly true in oncological treatment of pets, where cost is often no obstacle to applying all human treatment modalities to sick animals. As early as 1982, clients at the Colorado State University Cancer Center, the best in the world, were spending well over $100,000 on treating their pet in heroic ways. The question arises, from the animal's perspective, is the chance of extra life worth the significant additional suffering to the animal? This in turn leads to an ancillary question – can an animal value life per se? To answer this question we must consider some conceptual differences between animal and human cognition.

Human cognition is such that we can value long-term future goals and endure short-run negative experiences for the sake of achieving them. Examples are plentiful. Many of us undergo voluntary food restriction, and the unpleasant experience attendant in its wake, for the sake of lowering blood pressure or looking good in a bathing suit as summer approaches. We memorize volumes of boring material for the sake of gaining admission to veterinary or medical school. We endure the excruciating pain of cosmetic surgery to look better. And we similarly endure chemotherapy, radiation, dialysis, physical therapy, and transplant surgeries to achieve a longer, better quality of life than we would have without it or, in some cases, merely to prolong life to see our children graduate, to complete an opus, or fulfil some other goal.

In the case of animals, however, there is no evidence, either empirical or conceptual, that they have the capability to weigh future benefits or possibilities against current misery. To entertain the belief that 'my current pain and distress, resulting from the nausea of chemotherapy or some highly invasive surgery, will be offset by the possibility of an indefinite amount of future time,' is taken to be axiomatic of human thinking. But reflection reveals that such thinking requires some complex cognitive machinery. For example, one needs temporal and abstract concepts, such as possible future times and the ability to compare them; a concept of death, as we earlier saw as eloquently defined by Heidegger (1996) as 'grasping the possibility of the impossibility of your being'; the ability to articulate possible suffering; and so on. This, in turn, requires the possibility to think in an if-then hypothetical and counterfactual mode; that is, if I do not do X, then Y will occur. This mode of thinking, in turn, seems to necessitate or require the ability to process symbols and combine them according to rules of syntax in language. I do not deny that animals possess some (many) concepts such as food, water, conspecific, etc. but do affirm that they lack concepts based in complex syntactical constructions. Hence, for example, it is impossible to communicate negative states of affairs to animals, as argued earlier.

I would argue that the trapped coyote who chews off its leg does not do so because it anticipates consciously the possible consequences of immobility, but out of a far more primitive, hard-wired fear of being trapped, and correlative loss of control.

I have argued vigorously elsewhere against the Cartesian idea that animals lack thought and are simply robotic machines. It is also clear that animals have some concept of enduring objects, causality, and limited future possibilities (probably learned by association), or else the dog would not expect to get fed, the cat would not await the mouse outside of its mouse hole, and the lion could not intercept the gazelle. Animals also clearly display a full range of emotions, as Darwin famously argued (Darwin, 1872).

But it is also equally evident that an animal cannot weigh being treated for cancer against the suffering treatment entails, cannot affirm a desire (or even conceive of a desire) to endure current suffering for the sake of future life, cannot understand that current suffering may be counter-balanced by future life, and cannot choose to lose a limb to preclude metastases. We choose for them, ideally on the basis of clear benefit outweighing cost, as when we chose surgery and a few weeks of pain to treat animal cancer to assure additional years of life.

To treat animals morally and with respect, we need to consider their mentational limits. Paramount in importance is the extreme unlikelihood that they can understand the concepts of life and death in themselves rather than the pains and pleasure associated with life or death. To the animal mind, in a real sense there is only quality of life, that is whether its experiential content is pleasant or unpleasant in all of the modes it is capable of, for example whether they are bored or stimulated, fearful or not fearful, lonely or enjoying companionship, in pain or not, hungry or not, or thirsty or not. We have no reason to believe that, lacking linguistic tools, an animal can grasp the notion of extended life, let alone choose to trade current suffering for it. Obviously, they can value things that *entail* being alive, such as catching prey, without knowing the concept of life, or valuing life *per se.*

This, in turn, entails that we realistically assess what they are experiencing. We must remember, for example, that an animal *is* its pain, for it is incapable of anticipating or even hoping for cessation of that pain. Thus, when we are confronted with life-threatening illnesses that afflict our pet animals, it is not axiomatic that they be treated at whatever qualitative, experiential cost that may entail. The owner may consider the suffering a treatment modality entails a small price for extra life, but the animal neither values nor comprehends extra life, let alone the trade-off this entails (Rollin, 2006).

In the case of a laboratory animal, even one not experiencing constant pain, most of its *telos* is regularly violated, hence it is not happy and painless death is no harm to it. Food animals are a more interesting case. When they were historically raised under husbandry conditions accommodating their nature, they could be viewed as, at least to a significant extent happy – e.g. cows or sheep or pigs under pastoral conditions, without abuse. Under confinement conditions, virtually none of their *telos* is respected, so slaughter, especially when done in a minimally painful and minimally stressful way, without creating fear, is not only not a major source of harm, but a way of ending suffering and misery.

A very important corollary emerges from our discussion. We have argued that animals have no concept of death (or life) and consequently cannot value it more than pain. We have also indicated that people sometimes value death over pain, as a way of ending pain. If this is true of humans, it would be a fortiori true of animals, who cannot directly value life at all. Thus, in a sense, pain may well be worse for animals than for humans, as they cannot rationalize its acceptance by appeal to future life without pain. As I have said in other writings, a traditional argument affirms that human pain is worse than animal pain because humans can anticipate and fear pain very imaginatively before it happens, as when we plan to visit the dentist. Aside from the fact that animals too can fear imminent pain (e.g. when they cringe before a threatening upraised hand), the same logic decrees that animals cannot look forward to a time without the pain; their entire universe is the pain, they can have *no hope!*

4.3 Death and violation of *telos*

Thus far, we have argued that violation of *telos* can be worse for an animal than physical pain, as illustrated by the trapped coyote. We have also seen that pain can be worse than death, as when humans request death to end pain and the degradation, humiliation, and dependency that accompany uncontrollable pain. I would now argue that violation of *telos*, at least to a significant degree, can also be worse than death. Let us imagine an animal whose *telos* is well understood, for example a horse. A horse is a social herd animal whose nature involves running free. Let us further postulate that we now have confined a horse all the time in a small indoor enclosure, unable to run or socialize with other horses. Regrettably, we kept zoo animals under such conditions, until very recently and still keep chicken and pigs bred for food so severely restricted, though society appears is in the process of rejecting such management. Furthermore, let us suppose that we have totally eliminated the animal's ability to feel any pain at all. Unable to run, socialize, graze, choose its diet, be outdoors or fulfil any aspect of its *telos*, most of us would realize that the animal was not happy, indeed was miserable, despite the inability to feel pain. While lacking any concept of death, and thence being unable to fear death, the animal would certainly be aware of its inability to perform all aspects of its nature. Any harm to the animal in this case follows from its inability to exercise any portion of its *telos*. In such a situation, it would be monstrous *not* to euthanize the animal, as it has no functional life, merely protoplasmic endurance. Thus, providing such an animal with a gentle, painless death would in no way be inflicting harm upon it – quite the contrary, as one would ablate the frustration and misery of the animal's not being able to do anything constitutive of a horse's nature. Obviously, there is a much better alternative – simply change the conditions under which the animal is kept!

Such a situation is what is truly evil about industrialized confinement agriculture. It is not simply or primarily the fact that animals under these conditions feel pain, as sows certainly do when full-time life on concrete causes their legs to break down, and also causes pressure sores and muscle atrophy. It is not that they experience a greatly shortened lifespan. It is rather that their 'life' actualizes absolutely nothing of their nature. The ability to forage, root, choose what they eat, build nests, position those nests so that urine and faecal material run off, socialize with other sows, mate normally, move freely, play, cool off in mud, etc. Everything they are built and programmed to do is aborted and frustrated. And, lest anyone think that these animals do not know what they are missing, let us recall that my colleague, Temple Grandin, once released 20th generation pigs, bred for confinement, into an extensive environment, and they immediately went directly to a mud wallow (Temple Grandin, personal communication on multiple occasions).

Even when we vector respect for *telos* into making decisions regarding euthanasia of companion animals, we are not absolved of making difficult value judgments. Whereas at one time in the US, and in other countries today, people still view a three legged dog as incapable of fulfilling its nature, many citizens now realize the dog can still run, catch balls, chase rabbits, and generally actualize its *telos*. But in making euthanasia decisions, we must make value judgments not only concerning the amount of pain an animal is experiencing and for how long, but also regarding the degree to which it is capable of fulfilling its nature. In the case of companion animals, their *telos* will reflect not only the dictates of biology, but also the degree to which interactions with humans have shaped that nature.

References

Bentham, J., 1789. An introduction to the principles of morals and legislation. 2nd edition, 1823, chapter 17, footnote 122. Clarendon Press, Oxford, UK.

Darwin, C., 1872. The expression of emotion in man and animals. John Murray, London, UK.

Heidegger, M., 1996. Being and time (translator Joan Stambaugh). State University of New York Press, Albany, NY, USA.

Hume, D., 1779. Dialogues concerning natural religion. Second edition, London, UK.

Kilgour, R., 1978. The application of animal behaviour and the humane care of farm animals. Journal of Animal Science 46: 478-1476.

Mason, J., 1971. A re-evaluation of the concept of 'non-specificity' in stress theory. Journal of Psychiatric Research 8: 323-331.

Rollin, B., 1989. The unheeded cry: animal consciousness, animal pain and science. Oxford University Press, Oxford, UK.

Rollin, B., 2006. Animal rights and human morality, 3rd edition. Prometheus Books, Buffalo, NY, USA.

Weiss, J., 1972. Psychological factors in stress and disease. Scientific American 226: 101-113.

5. Do animals have a moral right to life? Bioethical challenges to Kant's indirect duty debate and the question of animal killing

H. Baranzke

Bergische Universität Wuppertal, Fakultät für Geistes- und Kulturwissenschaften - Kath. Theologie, Gaußstr. 20, 42119 Wuppertal, Germany; heike.baranzke@t-online.de

Abstract

Reflecting ethically on the end of animal life implies asking whether there is a duty to refrain from animal killing or whether there is a moral right to life for animals. From a Kantian point of view these questions are linked to the vivid philosophical debate about indirect duties with regard to animals and the doctrine of the duty-rights-symmetry. These doctrines lead to the core of Kant's ethical theory. Therefore, the indirect-duties-to-animals doctrine is extensively analysed in the context of the 'Doctrine of Virtues' of the 'Metaphysics of Morals' in order to meet three basic animal ethical concerns: whether it can include animals into moral considerations, whether it can consider animals morally for their own sake and not only for human advantages, and whether the animals' pain and suffering do count morally. Crucial with regard to the last aspect is Kant's concept of shared 'animality'. After this detailed elaboration of the dimensions of Kant's perfect duties to oneself with regard to refraining from maltreating animals, the results are questioned whether such a perfect duty to oneself is possible without exceeding Kantian ethical grounds, although Kant himself has considered the human being as being authorized to kill animals, when done quickly and painless. I shall show that such a *prima facie* duty is not only necessary for an integrative bioethical approach that consistently reflects upon human and animal needs, but even possible on the systematic grounds of a Kantian ethics. Nevertheless, there is no moral right to life for animals.

Keywords: animal ethics; theory of obligation; duty-right-symmetry

5.1 Introduction

Reflecting ethically on the end of animal life means dealing with the question whether killing animals is forbidden or whether there is a duty to keep animals alive when they are sick or hurt, and how these kinds of duties can be founded. If humans have a duty to refrain from animal killing, does this also mean that animals do have a right to life? Do animals possess a moral right to life like human beings do? Since Schopenhauer animal ethicists have complained about Kant's harsh denial and have, according to Robert Nozick, led to the assumption that Utilitarianism fits better to animal ethics than the Kantian ethics, which would work only in the field of human ethics. But a mere coexistence of different ethical theories remains unsatisfactory with regard to the needs of a bioethical approach being able to unify human and non-human bioethical affairs in one single consistent ethical theory. Therefore it is useful that recently a series of Kantian philosophers have tried to figure out how Kant's ethics could meet animal ethical challenges.

Though there is duty to abstain from causing needless pain to animals, a duty to refrain from animal killing does not exist in Kant's argumentation. Indeed Kant has explicitly allowed a quick painless killing of animals. Can a prohibition of animal killing easily be inserted into Kant's ethical approach? How could this work? And is there a significant difference between a possibility of a human's duty to refrain from needless animal killing and a possibility of an animal's moral right to life? And in case there is a difference, what would be the practical ethical implications?

Questions like these link the indirect duty debate to the problem of the duty-rights-symmetry. 'Indirect duties' denotes duties 'with regard to animals', which means animals cannot directly be addressed by human duties. Kant has treated these problems in the 'Metaphysics of morals', where he has maintained that humans can have duties only to human beings because non-rational beings are not capable of obligation. This statement has challenged animal ethicists and animal rights philosophers objecting to it by using the marginal case argument. It compares animals to morally incompetent human beings to unify both in the concept of a 'moral patient', which once more indicates the need of a consistent integrative bioethical theory. In order to meet this complex constellation of problems, it is decisive to understand the nature of Kant's indirect duty theory regarding animals, he has elaborated within the 'Episodic Section. On the amphiboly of moral concepts of reflection' in the 'Doctrine of virtue'. But the indirect duty theory cannot be adequately conceived without knowing the broader context of Kant's theory of duties. While doing this the intuitions of Kant's critics will be taken into account in order to figure out the underlying ethical problems, potential misunderstandings, and blind spots on this animal ethical battlefield (5.2). Against this background I shall raise the question of the morality of animal killing on a Kantian basis in order to profile a duty to refrain from unnecessary animal killing in contrast to a theory of a moral right to life restricted to humans alone (5.3), before ending with some conclusions (5.4).

5.2 Kant and the nature of indirect duties[1]

In the 'Basis of morality' Arthur Schopenhauer has shown himself embarrassed about Kant's statement that 'man can have no duty to any beings except human' and that therefore only indirect duties to oneself regarding animals are possible (Schopenhauer, 1995: 95f). Schopenhauer's critique on indirect duties to animals is strongly connected to his radical denial of the existence of duties to oneself. The same is the case in Leonhard Nelson's approach (Nelson, 1949). So the animal ethical critique of Schopenhauer and Nelson has two foci, one on a special fundamental ethical problem, whether there are duties to oneself, the other on the applied ethical intuition, that there must be direct duties to animals, if normative claims about their treatment shall be raised successfully. It is hermeneutically recommendable to take into account that Kant is mainly interested in the first issue and that his animal ethical remarks are little more than an explanatory example for his primary concern.

5.2.1 About the context and the primary function of Kant's remarks about duties regarding animals

Kant has developed his argument about the impossibility of direct 'duties to animals' within his theory of duties. In his 'Lectures on Ethics' he has critically followed the traditional scheme of duties as it was given by Alexander Gottlieb Baumgarten's compendium in ethics. Baumgarten had differentiated (1) duties towards God; (2) duties to oneself; (3) duties to others; and (4) duties to animals and spirits. But Kant was not convinced of the usefulness of this traditional cosmologically and religiously based duty scheme. Instead, he has developed the criterion of the ability of obligation in order to get a classification of duties against the background of his critical turn to an epistemic subject. Changing the criterion has implied replacing the standpoint of a pre-critical uninvolved observer by the critical self-reflecting, constantly involved epistemical and ethical finite subject. The critical turn, implying secularizing the human consciousness (cf. Ingensiep, 1996), has altered the architecture of justification arguments, reflected in the 'Episodic Section. On an amphiboly in moral concepts of reflection' in the 'Doctrine of Virtue' in the 'Metaphysics of Morals'. By virtue of the critical turn to the human subject in epistemology and ethics self-responsible decision-making within the limits of an enlightened, but finite human reason, autonomy instead of heteronomy, is possible.

---

[1] Kant's ethical works are cited in the traditional way, by the volume and page number of the standard German edition, Kants Gesammelte Schriften, edited by the Royal Prussion (later German) Academy of Sciences (Berlin: George Reimer, later Walter de Gruyter & Co, 1900ff.), which are found in the margins of most translations.

English citations from Kantian works are drawn from Denis (2005), Heath and Schneewind (1997) and Gregor (1996).

The 'Episodic Section' consists of three paragraphs, §§ 16-18 (MM, Ak 6: 443-444)[2], in which Kant has examined duty relations with the help of three kinds of things, namely 'inanimate' 'beautiful' natural things like 'crystal formations' and plants, animals as the 'nonrational part of creation' (§ 17) earlier characterized as 'endowed with sensation and choice' (MM, Ak 6: 442), and God as an only intelligible being (§ 18). Before illustrating the critical turn in ethics by these concrete types of cases Kant opens the section with remembering the nature of the analysis of the ethical phenomenon of self-obligation with regard to the human being, already given in §§ 1-3 (MM, Ak 6: 417-418), that he now generally applies to the type of a duty relationship between human and non-human beings (§ 16). In this passage his new critical ethical criterion for an enlightened consciousness of obligation can be found. It goes as follows:

> As far as reason alone can judge, a human being has duties only to human beings (himself and others), since his duty to any subject is moral constraint by that subject's will. Hence the constraining (binding) subject must, *first*, be a person, and this person must, *second*, be given as an object of experience, since man is to strive for the end of this person's will and this can happen only in a relation to each other of two beings that exist (for a mere thought-entity cannot be the cause of any result in terms of ends). But from all our experience we know of no being other than a human being that would be capable of obligation (active or passive). A human being can therefore have no duty to any beings other than human beings; and if he thinks he has such duties, it is because of an *amphiboly* in his *concepts of reflection*, and his supposed duty to other beings is only a duty to himself. He is led to this misunderstanding by mistaking his duty *with regard to* other beings for a duty *to* those beings (MM, Ak 6: 442).

Kant identifies two conditions for being a possible direct addressee for a duty: an ethical and an epistemological one. The ethical one is to be 'a person', a being with a 'will' 'that would be capable of obligation'. The ethical presupposition is necessary, but not sufficient. For Kant it needs to be accompanied by the epistemological precondition to be a being that can be sensually experienced. Thus God, traditionally conceived as a person with a will, cannot be conceived as a partner for obligation, because his existence cannot be objectively verified. How could an agent be sure of duties given by an epistemologically uncertain legislator who cannot be criticized by reason? Today, there are terrible examples of humans who feel religiously obliged committing murder by immunising themselves against reasonable critique. They commit an amphiboly in their moral concepts of reflection by which they mislead themselves about their moral

---

[2] Ak: Akademie-Ausgabe, the German standard edition by the German Academy of Sciences of Immanuel Kant's Writings. GMM: Immanuel Kant, Groundwort for the Metaphysics of Morals. LE: Immanuel Kant, Lectures on Ethics. MM: Immanuel Kant, Metaphysics of Morals

self-responsibility. Nevertheless, the religiously raised Kant acknowledges religious duties, but turns around the justification structure according to his critical philosophy. After having judged autonomously about the moral reasonability of prescriptions and after having recognized them as duties, in § 18 Kant declares it a human's perfect duty to oneself to apply the idea of God to the consciousness of the moral law in ourselves, *as if* a morally good deity would have legislated it (MM, Ak 6: 443; see also 6: 439f).

In contrast to God (and the traditional spirits in the 'Lectures') minerals, plants, and animals are given as sensible entities, but fail the necessary practical precondition to have the capacity of obligation, 'active' as well as 'passive'. Especially about animals Kant has said in his 'Doctrine of rights', that they are 'beings lacking reason, which can neither bind us nor by which we can be bound' (MM, Ak 6: 241). Thus, Kant adopts the Roman law's systematic division into 'things' and 'persons' with regard to animals and refers to it in his ethics, too. Here we meet the point of the modern controversy about the so-called 'moral status' of animals.

5.2.2 Person, dignity, end-in-itself – what animals are lacking by their nature

Influenced by an evolutionary view on the common origin of living beings and challenged by the biomedical relevance of the human being's own biological nature a growing number of people in the western industrialized part of the world feel uneasy by subsuming nature and especially higher animals under the traditional category of 'right in rem' that allows treating living beings as possession of persons. However, the Roman law tradition offers only two categories, things and persons, as a complete disjunction. In law an entity can only be a person or a thing, neither both nor is there a third alternative. It is not surprising that Kant's philosophy of law rests on the Roman law tradition. He links it to another couple of concepts, namely 'means' and 'ends' in his ethics. The concept 'means vs ends' derives from the Aristotelian teleology of action theory. It serves in Aristotle's ethics of goods to figure out which good is the final end or intrinsic good that cannot be used again as a means good for getting other goods, but is a good in itself. This good is called 'eudaimonia', happiness, and is only accessible to rational individuals like gods and humans. For Aristotle non-rational individuals like animals or plants are not able to take part in 'eudaimonia'. The Aristotelian 'eudaimonia' is *objectively* defined from an ideal observer's point of view in the frame of his cosmological teleology. In Kant's conception 'eudaimonia' is determined *subjectively* from the critical perspective of an inevitably involved, finite, sensuous subject of experience and agency. But both philosophers coincide that only rational virtuous beings, persons, are gifted to realize the only moral good and final end of a good rational will. Therefore according to Kant only humans possess 'dignity', which qualifies them as 'ends in themselves' that have to be respected as beings endowed with this specific moral nature to realize the moral good in the non-moral world. Their 'dignity' is priceless, since they cannot be replaced by other entities in the world, since these are not endowed with the ability of striving for realizing morality as the final end and objectively highest good. In so far we experience only human beings as being able to strive for realizing morality, called 'idea of humanity'. According to the 'humanity' or 'means-end' formula of the categorical imperative the 'idea of humanity' shall be treated 'whether in your person or in that of any other, in every case at the same time as an end, never as a means only' (cf. GMM, Ak 4: 429). That means, that we always should be aware of other human beings who are, like us, endowed with the destination of moral agency. Obviously, Kant's notion of humanity goes far beyond the biological characteristic of being a member of the species *homo sapiens sapiens* (cf. Wood 1998: 189). The 'idea of humanity' is not defineable biologically. Thus, animal ethical approaches (like Korsgaard, 2004) that try to integrate non-moral entities like animals into the means-end-formula of the categorical imperative is condemned to fail (cf. Denis 2005: 87 n. 1). They fail even on Aristotelian grounds, because Aristotle's and Kant's ethics both coincide in considering the final end as a *moral good* and not a *natural good*. Even Aristotle distinguishes 'eudaimonia' as a practical term, applicable only to rational beings, from 'entelecheia' as a theoretical term of natural philosophy. From a modern ethical standpoint animal ethical argumentations, which miss this difference and try to apply dignity or intrinsic value to animals, can be reproached to commit Hume's is-ought-fallacy. They declare animals as morally valuable due to their biological characteristics without justifying this adscription morally.[3]

Morality can only be experienced from a first person's moral perspective through the observation of oneself as a moral agent that can be reflected and analysed transcendentally. The transcendental ethical method asks for the inevitable *conceptional* (not empirical!) conditions of the possibility of acting morally. In § 11 of his 'Doctrine of virtue' Kant has analysed what we are inevitably doing when making moral claims. And we cannot escape from making claims as human beings. We demand respect from others, who we – by doing this – discover as 'equals' who can demand the same from us (MM, Ak 6: 435). By mutually fostering respect we presuppose each other as beings being able to be obligated by us and discover ourselves as beings, who can be obligated by another will.

Day by day we experience this moral phenomenon, e.g. when making a date and expecting that our dating partner will meet us as arranged. Further, we are aware that we would never confront an animal with this moral claim. However, we can hardly avoid meeting human beings with this sort of moral expectation, otherwise we would not notice them as human beings. In the case of little children we can rationalize that they still are not able to perform this moral capacity. But we raise our children in the horizon that one day they will be capable moral agents and will ask us how we have

---

[3] In Europe, since the year 1992, the debate is additionally stimulated by the Swiss case of the new constitutional concept 'Würde der Kreatur', which is usually translated into 'dignity of creatures' (cf. Baranzke, 2012).

treated them in earlier times. In the case of adults who do not perform the expected moral behaviour, it depends on whether we believe that they are mentally disabled or not willed to respond to our moral expectations, if we react to their behaviour with regret or anger. These kinds of usual reactions indicate that humans meet one another inevitably in the horizon of moral expectations, because they are human beings. In Kant's language humans meet one another as equal inhabitants of the 'kingdom of ends' (GMM, Ak 4: 435). Thus, dignity in the sense of being bound to act morally is an 'inner worth' (MM, Ak 6: 418) that cannot be graduated by observable measurements of biological or psychological data. The popular marginal case argument that compares humans and animals on the grounds of data-based empirical characteristics is misleading, since it presupposes a merely biological concept of the human being while ignoring the moral horizon of claiming mutual respect (cf. also Kain 2010: 34). Avishai Margalit has described this way of undignifying human beings as treating humans *as if* (!) they were no humans in his book 'A decent society' (1996). According to Margalit somebody who, in fact, does not realize a human being as a human being, is a mentally disabled person who needs to be treated by a physician, like 'The man who mistook his wife for a hat' (1985) in Oliver Sacks' clinical tale.

5.2.3 Kant's transcendental anthropology of obligation and its function in a theory of duties

But how is obligation possible? Kant has elaborated an explanation theory in §§ 1-3 of the 'Doctrine of virtues', which opens his 'Ethical elementary doctrine' by laying the transcendental foundations for duties to oneself (Mozayebi, 2013). While § 11 shows *that* human beings intuitively conceive themselves and other human beings as addressees of obligation, §§ 1-3 demonstrates *how* a human being may conceive herself to recognize herself as being able to self-obligation. Such a theory is important, because a concept of self-obligation is the systematic condition for the possibility of having duties, namely either duties to oneself or duties to others. Then: 'suppose there were no such duties: then there would be no duties whatsoever, and so no external duties either. For I can recognize that I am under obligation to others only insofar as I at the same time put myself under obligation, since the law by virtue of which I regard myself as being under obligation proceeds in every case from my own practical reason' (MM, Ak 6: 417f). The transcendental self-reflective theory of moral agency makes autonomous self-obligation plausible for a living entity with a mixed sensual and rational nature. For this purpose the human being has to be regarded in two perspectives, namely as a 'homo phaenomenon' in the passive role of the *subiectum obligationis* and as a practical 'homo noumenon', the idea of humanity, in the active role of the 'auctor obligationis'. The true addressee for having a 'duty towards' is the noumenal idea of humanity, the 'personality' (MM, Ak 6: 418), in a given phenomenal person.

The outcome of these metaethical fundaments concerning animals is that they cannot be conceived as beings capable to oblige or be obligated. They cannot be considered as addressees for duties. Thus, in a theory of duties animals can only be reflected indirectly as objects of human duties regarding them, as matters of human moral self-obligation. But the fact that 'animals are not themselves *owed* moral consideration' (O'Hagan, 2009: 534) does not imply that it is impossible 'to include animals' (O'Hagan, 2009: 532) into moral concern 'for their own sake' and 'not as mere means, having only an extrinsic and instrumental value' (Wood, 1998: 194), as often assumed. In contrast to these worries the concept of indirect duties with regard to animals succeeds in integrating animals for their own sake into a duty theory without neglecting their incapability of being obligated. It is remarkable that animal ethicists usually are not interested in a theory of obligation, although it is of great importance for their genuine aims. They take obligation as an evident moral phenomenon. In so far animal ethical approaches are in danger to be based upon weak ethical groundings that cannot supply the raised animal ethical claims. Animal ethics like all kinds of applied ethics should consistently be based on a philosophically sound theory of obligation that can explain, how obligation is possible for human beings as the only possible addressees who can understand claims of animal ethicists.

5.2.4 The non-moral nature of animals and their role in a duty theory

Kant has used his transcendental anthropology, presupposed in § 16 MM, in order to subdivide Part I of his 'Ethical elementary doctrine' into the two categories of perfect duties to oneself (Book 1) and imperfect duties to oneself (Book 2). Part II deals with 'The virtual duties towards others' – other human beings are meant. In the first main section of Book 1 Kant develops the *perfect* inner *duties* of the human being to oneself regarded as a phenomenal being, where we find strict prohibitions of abusing one's own person by e.g. committing suicide. In the second main section of Book 1 he refers to the human being's perfect duties to oneself, regarded as a purely moral being, where fundamental issues like self-judgement, moral consciousness, and human dignity are reflected. In the *second book* about the *imperfect duties to oneself* duties of cultivating one's talents (with regard to the human being as 'homo phaenomenon') and duties of working on one's own moral perfectibility (with regard to the human being as 'homo noumenon') are dealt with. Regarding our main concern – the human-animal relationship – it is significant, that Kant has placed the 'Episodic Section' at the end of Book 1 in the first part, completing the treatment of the perfect duties to oneself and before referring to the imperfect duties to oneself and the perfect and imperfect duties to others.

Because animals cannot be conceived as beings capable to oblige or be obligated at any time in their existence, they are, in contrast to the human being, non-moral entities, neither 'morally innocent' (cf. Timmermann, 2005: 145 note 24) nor able to become guilty. Applying moral concepts, like 'innocence', 'guilt', 'obligation', 'duty', 'right', etc. to non-moral entities like animals does not make any sense ethically, but reveals a form of uncritical anthropomorphization of animals. But claiming that 'another animal can obligate you in exactly the same way another person can' (Korsgaard 1996: 153) means committing an amphiboly in the concept of moral reflection or a 'subject-addressee fallacy', as Otfried Höffe (1993: 215) has called it. Every day we confirm that animals cannot be thought as endowed with a free will and a noumenal nature by which they were dignified to be obligated to realize the moral law, since we do not try seriously to make dates or contracts with animals. In the traditional philosophical language of the 18th century it sounds harsh when reading that animals cannot be respected as ends in themselves and cannot be included in the end-means-formula of the categorical imperative because they are only means for human beings to reach moral ends. Nevertheless, a growing number of even critical Kant interpreters has to concede that his ethics works, at least pragmatically regarded, not worse than those of his rivals in the field (Baranzke, 2002; Denis, 2000; Kain, 2010; O'Hagan, 2009; Timmermann, 2005).

There is, however, some confusion about the kinds of moral ends non-moral animals shall serve for as means. Should the answer be derived from the duty to oneself under which Kant has subsumed his section about animal treatment in the 'Doctrine of virtues' or should we follow the reasons Kant has delivered for his prohibition of cruelty to animals? Many animal ethicists have been convinced that maltreating animals is forbidden for the sake of other human beings (e.g. Regan, 2004: 179; Timmermann, 2005: 139), 'for it dulls his shared feeling of their suffering and so weakens and gradually uproots a natural predisposition that is very serviceable to morality in one's relations with other men' (Ak 6: 443). 'Thus only for practice are we to have sympathy for animals, and they are, so to speak, the pathological phantom for the purpose of practicing sympathy for human beings', as Schopenhauer commented in the 'Basis of morality' (1995: 95f).

However, the so-called cruelty or brutalization account, which can be traced back to ancient times, is a hypothetical empirical argument that may be true or not. Despite some plausibility for in the long run cruel behaviour will brutalize people's behaviour to human beings, the argument cannot reliably found a strict prohibition (cf. Baranzke, 2004; Kain, 2010; O'Hagan, 2009). Allan Wood has driven the argument into absurdity by imagining that 'if it happened to be a quirk of human psychology that torturing animals would make us that much kinder toward humans ..., then Kant's argument would apparently make it a duty to inflict gratuitous cruelty on puppies and kittens so as to make us that much kinder to people' (Wood, 1998: 194). Fortunately, neither human psychology seems to be a quirk, nor did Kant make his prohibition to maltreat animals dependent from empirical grounds. In the 'Lectures on ethics' Kant has characterized the argumentation as 'a good lesson to children' (LE, Ak 27:

459), nothing more. It became very popular in the second half of the 18th century on behalf of the engraving cycle 'Four Stages of Cruelty' by the British artist William Hogarth (Kottenkamp, 1840) published in 1751, to whom Kant explicitly refers. This may explain why he could not have resisted indorsing it into § 17 of the 'Doctrine of Virtue' for its pedagogical attractiveness.

While usually most animal ethicists complain about Kant's instrumentalization of animals for (other) human's sake, Lara Denis (2000) acknowledges the broad range of positive moral and psychological effects refraining from animal maltreatment has. But both approaches neglect Kant's decisive reason why besides the range of possible effects animals shall be taken into moral account, namely, that refraining from animal maltreatment and even gratitude for their services done over years belong to 'man's duty *to himself*' (MM, Ak 6: 443). Kant does not qualify his founding as 'additionally besides' the other duty category, the duties *to other human beings*. Moreover 'man's duty *to himself*' is the only category he explicitly offers.

But, as Svoboda (2012: 154) recently has underlined, Kant 'does not identify explicitly the duty to oneself upon which duties regarding non-humans depend'. As I have pointed out in earlier writings (Baranzke, 2002, 2004, 2005), it seems to be not just by accident that Kant has placed the 'Episodic Section' as a concluding section of Book 1 about the perfect duties to oneself, before he has proceeded his examinations on the imperfect duties to oneself and the duties to others (cf. Kain, 2010: 221). Perfect duties to oneself are characterized as 'negative duties', which '*forbid* a human being to act contrary to the *end* of his nature and so [they] have to do merely with his moral *self-preservation*' (MM, Ak 6: 419). 'Moral self-preservation' has to be distinguished from 'moral self-perfection'. The latter is an imperfect wide duty to strive for one's own moral cultivation (cf. Kain, 2010: 222). So, perfect duties to oneself are not duties of perfectibility. Kant has surely not considered duties regarding animals as 'part of the perfection of ourselves as natural ... beings' (O'Hagan, 2009: 534). Although in the 'Lectures on ethics' caring for animals is described as to 'cultivate my duty to humanity' (LE, Ak 27: 459), it is not convincing to subsume refraining from animal maltreatment and the 'wanton destruction' of inanimate nature (MM, Ak 6: 443) only under the imperfect duties to oneself as duties to oneself to increase one's own moral perfection, like Allen Wood (1998: 195) and Toby Svoboda have argued, 'because they [scil.: such actions] do quite the opposite' (Svoboda, 2012: 159), as Svoboda has noticed himself. Further, it is not convincing because cultivating duties always refer to the empirical psycho-physical nature of the moral agent. Thus imperfect duties to increase physical talents or psycho-moral inclinations would be affected by the same objections concerning being only a hypothetical argument that can be falsibilized like the cruelty-account. I think part of the problem concerning imperfect duties is that they oscillate between the (hypothetical) empirical internal effects on the agent herself, and the question about the external executive degree of an action, the estimation about how much has to be done.

After all, in contrast to Svoboda's proposal (2012: 155) it seems more plausible to consider animal torturing like lying, avarice, and servility (MM, Ak 6: 429-437) as acts that hurt the agent's 'moral self-preservation' without any internal or external relevance of degree. Only gratitude concerning the long service of working animals is a candidate for an imperfect duty to oneself, in so far obeying 'the *intensity* of gratitude, that is, the degree of obligation to this virtue, is to be assessed' in order 'to cultivate one's love of human beings' (MM, Ak 6: 456), Kant explains with regard to other human beings (§ 33). But gratitude with regard to animals is just a minor aspect in § 17 of the 'Doctrine of virtues' – Svoboda himself has alluded to §§ 21-22 of the 'Doctrine of Virtues'. But the duty of '*purity* of one's disposition to duty' and the duty of striving for fulfilling all duties are very general duties on a meta-level, which can be applied to all specified duties. They are, as Kant explains, '*narrow* and perfect ... in terms of its [scil.: the duty to oneself's] quality', but 'wide and imperfect in terms of its degree, because of the *frailty* (*fragilitas*) of human nature' (MM, Ak 6: 446). Here, Kant reflects quite generally on impurity and finiteness of the human being as a moral agent with regard to the *degree of succeeding* in the moral execution of duties, but not with regard to the *reason of obligation*, which is '*narrow* and perfect'. In consequence, I agree with Kain who objects to Timmermann and Wood that 'proper treatment of animals' can be regarded as 'a necessary condition for and perhaps a constitutive part of one's moral well-being, rather than a mere 'instrumental' means to it'. Focussing 'on the agent's self-respect' in a way, as if Kant's account would 'foreground the agent's self-concern ... and background her concern for the animal' is correctly characterized as 'psychologically peculiar and ethically deficient' (Kain, 2010: 227). It is odd to claim *moral* consideration of animals, while simultaneously assessing animals would be instrumentalized for the sake of *morality*.

Kant's subtle argumentation in § 17 and his composition of the 'Episodic Section' at the end of Book 1 in his duty taxonomy proof duties with regard to animals as being *perfect* moral duties to oneself, by which Kant has intended no other purpose than realizing a strict moral end, namely to prohibit animal torture, painful killing, needless vivisection, and straining working animals beyond their capacities. Supposing that 'only human interests matter morally because they are the interests of human beings' (Cohen and Regan, 2001: 286), is surely not true for Kant's ethical argumentation regarding animals. Despite it is a strict reason of obligation why animal maltreatment is forbidden and therefore a perfect duty regarding animals towards the moral agent herself, one question is still unanswered, namely the question about the ethical role of the well-being of living beings.

5.2.5 Shared animality and the question of the nature of its moral significance

Although it can be demonstrated that the concept of indirect *duties with regard to* animals includes animals into moral consideration despite of their lacking the capability of obligation, and although it could be proved that *perfect duties to oneself* regarding animals exclude all kinds of instrumentalizing animals for human advantages, a further concern has to be met: Animal ethicists doubt whether *indirect* duties can capture 'that an animals' pain is directly morally significant' (Timmermann, 2005: 139), that 'animal pain matters directly' (O'Hagan, 2009: 542). 'But why don't we owe these duties directly to the other animals?' (Korsgaard, 2004: 91).

This third pathocentric challenge of the indirect duty approach can be objected by remembering that Kant has criticized the Cartesian animal machine theory (Naragon, 1990), because he has taken the nature of animals as sentient beings seriously. In § 17 Kant refers explicitly to '*their* suffering' (MM, Ak 6: 443) by strictly prohibiting maltreating animals. It seems a lucky English translation to me to speak of 'shared feelings' ('Mitgefühl') for it underlines the common 'animality' of humans and animals with regard to their sensitive nature, which Lara Denis has reflected in her contribution to 'Kant's conception of duties regarding animals' (2000). Also Kant's monition, not to use animals beyond their limited capacities, is a further proof for his concern for their vulnerability, because of which in his 'Lectures' Kant has characterized animals as 'analogue of humanity' (LE, Ak 27: 459). It seems that in his lessons Kant has made use of the traditional concept of *analogy* in order to emphasize the empirical *similarity* between animals and humans on behalf of their vulnerable and sensitive nature and, *at the same time*, their *decisive difference* concerning their transcendental moral agency. In his authorized publications we find a preference for the opposition of 'animality' ('Tierheit') vs 'humanity', respectively, 'personality'. The double determination of the human-animal-relation, transcendental moral difference and empirical similarity, poses the question how ethical difference and empirical similarity relate to each other with regard to the duties they meet: the duties of respect and the duties of love.

In fact, Kant has even applied gratitude as an imperfect 'duty of love' to animals (Denis, 2000: 409; Baranzke, 2004: 8). But he could not address animals directly like other human beings, since animals cannot be ethically addressed, neither active (*auctor obligationis*) nor passive (*subiectum obligationis*). Therefore, the reason for obligating strictly to be concerned about animal suffering and even animal well-being is the idea of humanity in the moral agent's person. So it is a *perfect* duty to oneself in terms of the *obligating force* (Verpflichtungsgrund). But it is an *imperfect* duty with regard to the matter of responsibility (Verpflichtungsgegenstand), that is *empirical well-being*, and with regard to the question to what *degree* or extent well-being of other beings, animals as well as humans, has to *be promoted*. But *that* considering animal well-being as a duty derives from the fact, that analogously well-being is relevant for finite and vulnerable human beings, too. Thus, negative duties of respect need to consider complementarily the empirical finiteness and vulnerability of human beings as 'fellowmen, that is, rational beings with needs' (MM, Ak 6: 453) in order to come to concrete moral judgements. From this Korsgaard draws the conclusion: 'It is therefore our animal nature, not just our autonomous nature, that we take to be an end-in-itself' (Korsgaard, 2004: 104) With this argumentation of shared animality as a shared 'natural good' she intends to pave the way to show animals as ends-in-themselves, too. However, due to the theory of obligation, 'it is still inconceivable' that a human being 'should have a duty to a *body* (as a subject imposing obligation), even to a human body' (MM, Ak 6: 419), because bodies and their empirical characters as well as desires or feelings of uneasiness are merely empirical facts. Using them ethically as moral reasons means committing Hume's fact-value fallacy. Instead, they have to be judged morally with the help of non-empirical ethical criteria. However, empirical features are important pieces of epistemic information about the relevant nature of an entity. They help a moral agent to judge adequately informed about *how* a given entity should be treated in a morally justifiable way – according to the best epistemic knowledge with regard to an entity's empirical nature and the best moral consciousness with regard to the reasonability of the agent's own reasons. So, empirical features play an important role in the process of concrete moral judgements about e.g. the special vulnerability and concrete needs of a living individual. But they cannot serve as moral reasons. On this background we recognize that Kant's 'animality' serves as a general denote of the shared biological nature of humans and animals, which has to be made more concrete with regard to practical judgements in special applied animal ethical cases.

5.2.6 Moral judgement and the question of 'good reasons'

The previous section has shown that well-being or suffering, are relevant empirical facts which have to be considered in concrete practical ethical judgements. In an ethical theory of morally relevant goods they can be reflected as empirical inner constraints and outer conditions, which are necessary to consider for finite and vulnerable beings to perform agency. But how and to what extend they have to be taken into consideration, has to be decided in a procedure of weighing up 'moral goods'. Therefore, applied to animals as the 'analogues to humanity', Lara Denis has spoken about '*prima facie* duties' regarding the well-being of animals (Denis, 2000: 409). Facing the moral good of well-being a virtuous moral agent has to ask himself sincerely whether there is really a morally good reason to make an exception from the *prima facie* duty to refrain from making an animal suffer. Further, a decent society can ask the same question and is free for obligating itself, even by positive laws, to acknowledge certain standards of animal treatments. Allen Wood has correctly underlined that although 'Kant mentions no specific juridical duties regarding animals or the natural environment … it is worth noting that there is room for them. In Kant's theory, the fact that nonrational beings have no rights does not entail that the general will of a state may not legislate restrictions on how they may be used or treated' (Wood, 1998: 192).

5.3 Enjoying being alive and the question of analogy

In his times, Kant has looked at vivisection as being a problematic case, which challenges the moral agent to proceed a concrete practical judgement, the enterprise of 'weighing up moral goods'. But Kant's own argumentation has to be questioned why the life of an animal shall not be regarded as a 'moral good'?

5.3.1 Is animal killing morally relevant on Kantian grounds?

Like the chief witness of modern animal ethics, Jeremy Bentham, or the meat eating philosopher of sympathy, Arthur Schopenhauer, Kant has had no doubts that humans were allowed to take the life of an animal if it is killed quickly without causing it long suffering (cf. § 17, MM, Ak 6:443). Despite we can go further on than Kant himself has done and can ask: What is a good reason for taking the life of an animal? We can ask this question, because animals are not only an 'analogy to mankind' with regard to their sensitive nature, but also with regard to their enjoying being alive. Although biologically being alive is not the moral final end, it is nevertheless a morally relevant empirical good, a natural condition for moral beings to perform morality. Therefore we have to respect being alive for moral ends. Analogously animals need to be alive in order to follow their own natural goods like to enjoy their life or to raise their offspring. There is no good reason for excluding the issue of being alive from the empirical analogy of the 'shared animality' between human beings and animals, although being alive is no end-in-itself neither for animals nor for humans. So, on the grounds of a Kantian ethics there is no obstacle for arguing that there is a perfect duty to oneself to refrain from animal killing without morally good reasons.

5.3.2 What is the difference between a perfect duty not to kill animals and an animal's moral right to life?

Nevertheless, a perfect duty to refrain from animal killing without a good reason is not the same as an animal's moral right to life. There is no analogy to the human right to life, like the proponents of the Great Ape-Project (Cavalieri and Singer, 1993) evoke, for animals cannot have duties and therefore cannot claim moral right. Applying moral terms like 'duty' or 'right' or 'claim' is meaningless for them, not only accidently for single individuals, but permanently and in general for all members of a species, as we have seen above. This is the sense of the doctrine of the duty-rights-symmetry, namely that semantically only those kinds of beings, that, in general, can be obligated, can have rights. The source of morality in animal ethics is the human being by virtue of his moral nature, because only humans can be considered as being able to oblige and be obligated. If there were no human beings able to obligate themselves, no duties or rights would exist at all in the world, no duties to regard animal ethical claims included. Therefore, claiming a moral or natural animal right to life is a result of the amphiboly in moral concepts of reflection. The proclamation of moral, respectively, natural animal rights seems to try to forget that it is always the human being who has to make a morally justifiable decision by means of the procedure of a practical judgment. It is noteworthy how the protagonist for animal rights, Tom Regan, at first has suggested a moral right to life for animals by replacing the criterion of moral agency through the empirical criterion of being a subject-of-a-life (Regan, 2004: 243-248). But finally, in his life boat case dilemma, he has decided to save the four humans and not the dog (Regan, 2004: 324). Regan cannot avoid distinguishing morally between human beings and animals, although maintaining their equal right to life based on their equal inherent value.

5.4 Outlook on the basis of moral considerations on the end of animal life in an integrative bioethics

Like most of his contemporaries Kant himself has not regarded animal killing, when done quickly and as painless as possible, as morally problematic. But the technically induced growing biomedical ethical concern about end of life decision raise the question why only human biological life especially at its margins should be considered so carefully. The inner logic of the bioethical end of life debates drives to the necessity of developing an integrative bioethical theory that allows considering human and animal bioethical affairs according to consistent ethical principles. Kant's ethics already supports this necessity with the concept of shared 'animality' that reflects the given empirical similarities between human and animal beings without declaring them as moral ends-in-themselves; this would step into the trap of Hume's is-ought-fallacy. Kant himself has only considered pain, suffering and some aspects of the well-being of working animals with the help of the concept of shared animality. But being alive belongs to the catalogue of empiric human-animal-similarities, too. Thus, there is no reason to exclude the question of killing animals from an ethical consideration on Kantian grounds. Bioethical thinking has made us more sensitive for the necessity to acknowledge humans and animals as 'analogues' also with regard to being alive.

A decisive advantage of a Kantian bioethical approach is that it can meet the challenges of a naturalistic anthropology and an undermining of the fundaments of a human rights philosophy. So Kantian ethics avoids fatal moral consequences of biologistic and reductionistic bioethical approaches. Furthermore, Kant's ethical theory about obligation as a moral phenomenon reveal the ethical presuppositions inevitably required by every human being and every ethicist by raising moral claims. Raising moral claims is not the same like having needs. It presupposes a moral self-understanding about the capacity of obligating and being obligated. Therefore humans and animals can only be considered as 'analogues' despite their shared animality.

Thus the task left over for every human being, in general, and e.g. for veterinarian professionals in more special cases is, to engage one's moral capacity for practical judgement to consider in which cases it may be allowed to make an exception from the prohibition of killing animals without reaching a situation of self-contradiction with regard to weigh morally on similar 'moral goods' or interests. But although it demonstrates an amphiboly on moral concepts of reflection to claim a natural or moral right to life for animals, the question of a 'good reason' that allows taking an animal's life is raised. It does not only challenge human individuals. Our societies are confronted with it, too. Positive animal protection laws and their consequent prosecution reflect how seriously morally this question is taken.

Acknowledgements

I thank the editors of this volume for useful comments on an earlier draft of the article and Hans Werner Ingensiep, University of Duisburg-Essen, for inspiring discussions.

References

Baranzke, H., 2002. Würde der Kreatur? Die Idee der Würde im Horizont der Bioethik. Würzburger Wissenschaftliche Schriften. Reihe Philosophie Band 328. Königshausen & Neumann, Würzburg, Germany.

Baranzke, H., 2004. Does animal suffering count for Kant? A contextual examination of § 17 in the Doctrine of Virtue. Essays in Philosophy 5(2): 4.

Baranzke, H., 2005. Tierethik, Tiernatur und Moralanthropologie im Kontext von § 17 Tugendlehre. Kant-Studien 96: 226-363.

Baranzke, H., 2012. Why should we make a difference? How dignity can work in animal ethics. Bioethica Forum 5: 4-12.

Cavalieri, P. and Singer, P. (eds.), 1993. The great ape project. Equality beyond humanity. Fourth Estate, London, UK.

Cohen, C. and Regan, T., 2001. The animal rights debate. Rowman & Littlefield Publishers, Inc., Lanham, MD, USA.

Denis, L., 2000. Kant' s conception of duties regarding animals: reconstructions and reconsideration. History of Philosophy Quarterly 17: 405-423.

Denis, L. (ed.), 2005. Immanuel Kant. Groundwork for the metaphysics of morals. Translated by Thomas K. Abbot with revisions by Lara Denis. Broadview Editions, Ltd., Ontario, Canada.

Gregor, M. (ed.), 1996. Immanuel Kant. The metaphysics of morals. Cambridge Texts in the History of Philosophy. Cambridge University Press, Cambridge, MA, USA.

Heath, P. and Schneewind, J.B. (eds.), 1997. Immanuel Kant. Lectures on ethics. The Cambridge Edition of the Works of Immanuel Kant. Cambridge University Press, Cambridge, UK.

Höffe, O., 1993. Moral als Preis der Moderne. Ein Versuch über Wissenschaft, Technik und Umwelt. Suhrkamp, Frankfurt am Main, Germany.

Ingensiep, H.W., 1996. Tierseele und tierethische Argumentation in der deutschen philosophischen Literatur des 18. Jahrhunderts. Internationale Zeitschrift für Geschichte und Ethik der Naturwissenschaften, Technik und Medizin N.S. 4: 103-118.

Kain, P., 2010. Duties regarding animals. In: Denis, L. (ed.) Kant's metaphysics of morals. A critical guide. Cambridge University Press, Cambridge, MA, USA, pp. 210-233.

Kant, I., 1900ff. Kant's Gesammelte Schriften. Akademie-Ausgabe (Ak). Georg Reimer, later Walter de Gruyter & Co, Berlin, Germany.

Korsgaard, C. (ed.), 1996. Sources of normativity. Cambridge University Press, Cambridge, UK.

Korsgaard, C., 2004. Fellow creatures: Kantian ethics and our duties to animals. In: Peterson, G.B. (ed.) The tanner lectures on human values. Salt Lake City, UT, USA, pp. 77-110.

Kottenkamp, F. (ed.), 1840. W. Hogarth's Zeichnungen, nach den Originalen in Stahl gestochen. Mit der vollständigen Erklärung derselben von G.C. Lichtenberg. Literatur-Comptoir, Stuttgart, Germany.

Margalit, A., 1996. The decent society. Harvard University Press, Cambridge, MA, USA.

Mozayebi, R., 2013. Die 'Antinomie' des § 3 der Tugendlehre. In: Bacin, S., Ferrari, A., La Rocca, C. and Ruffing, M. (eds.) Kant und die Philosophie in weltbürgerlicher Absicht. Akten des XI. Internationalen Kant-Kongresses 2010. De Gruyter, Berlin, Germany, pp. 443-455.

Naragon, S., 1990. Kant on Descartes and the Brute. Kant-Studien 81: 1-23.

Nelson, L., 1949. System der philosophischen Ethik und Pädagogik. Aus dem Nachlass hrsgg. v. Grete Hermann und Minna Specht, Göttingen, Germany.

O'Hagan, E., 2009. Animals, agency, and oblligation in Kantian ethics. Social Theory and Practice 35: 531-554.

Regan, T., 2004. The case for animal rights. University of California Press, Berkley, CA, USA.

Sacks, O., 1985. The man who mistook his wife for a hat. Gerald Duckworth & Co Ltd., London, UK.

Schopenhauer, A., 1995. On the basis of morality. Translated by E.F.J. Payne, with an introduction by David E. Carthright. Berghahn Books, Providence, RI, USA.

Svoboda, T., 2012. Duties regarding nature: a Kantian approach to environmental ethics. Kant-Yearbook 4: 143-163.

Timmermann, J., 2005. When the tail wags the dog: animal welfare and indirect duty in Kantian ethics. Kantian Review 10: 128-149.

Wood, A.W., 1998. Kant on duties regarding nonrational nature I. Supplement to the Proceedings of The Aristotelian Society 72: 189-210.

6.1 Introduction

The current debate on human-animal relations and animal ethics often focuses on pertinent inconsistencies (cf. Bulliet, 2005; Herzog, 2011; Joy, 2010). The most controversially debated Dutch artist Tinkebell puts her finger right on it.[1] She intended to kill 61 day-old chicks in a performance. When she announced what she was going to do, this was followed by a foreseeable and huge media response. Taking into account that over four billion day-old chicks are killed worldwide annually (cf. Aerts *et al.*, 2009: 117) with comparatively little media response, this outcry raises several questions. Tinkebell comes to the conclusion that our attitude towards animals is pathological and paradoxical. Treating (biologically) equal animals unequally is often described and considered contradictory and immoral (cf. Kunzmann, 2013: 57f; Singer, 2011). Those contradictions lead e.g. People for the Ethical Treatment of Animals to the catchy slogan: 'If your dog tasted like pig, would you eat him? What is the difference?'[2]

Why do we treat biological equals unequally? Especially when it comes to the issue of killing animals, these inconsistencies and contradictions appear to be particularly obvious: What is the 'substantial' difference between the rabbit in the living room, kept as a companion and euthanized on humanitarian grounds, the rabbit in the cage in the laboratory (probably waiting for a painful procedure in a lab) and the rabbit on the plate, served with sauce and dumplings? What are the crucial differences between the dog and the pig, the horse and the cow? Undeniably, they are similar in biological terms, but are treated in extremely different ways. Depending on the context they live in, different standards and practices of killing are applied. 'Being in the wrong box' serves as a proverb to indicate that we face severe inconsistencies in human-animal relationships.

This phenomenon can be addressed from at least two viewpoints. The first is revealed irrationality or inconsistency: if our morally relevant relations to animals are to be determined on the basis of biological or cognitive capacities only, common practices express an unequal treatment of equals that contradicts our idea of justice. This position resembles moral individualism, i.e. that our moral duties towards animals are determined by their (natural) capacities. According to moral individualism in animal ethics, the source and addressee of moral respect is the individual animal. If we want to know how animals should be treated, the answer lies 'in the animal':

> What distinguishes approaches in ethics that count as forms of moral
> individualism is the claim that a human or nonhuman creature calls for

---

[1] Süddeutsche Zeitung, 31 May 2009. Available at: http://tinyurl.com/33llsh2.

[2] This is our translation of the German version: 'Wenn Ihr Hund nach Schwein schmecken würde, würden sie ihn dann essen? Wo ist der Unterschied?' Available at: http://www.peta.de/web/anzeigen.2462.html.

specific forms of treatment only insofar as it has individual capacities such as, for instance, the capacity for suffering or the capacity to direct its own life (Crary, 2010: 20; cf. McMahan, 2002).

Peter Singer and Tom Regan can be named as the two most prominent moral individualists in animal ethics. In their view, the role of perspectivity and the human viewpoint should be set aside. We search and presume knowledge about the capacities and the natural existence of animals. And obviously, if different animal species and humans are similar concerning morally relevant characteristics, they have to be treated similarly for reasons of consistency.

What has been coined the extension model of moral respect by McReynolds, defines this line of argumentation for moral standing in Singer's and Regan's view well: 'Whenever moral standing is extended to a new group, it is granted to the new group to the extent of and on the basis of their similarity to members of the old group' (McReynolds, 2004: 64). The characteristics of an initial group of individuals with moral standing are used to extend the group on the basis of similarity as the criterion of judgment (cf. Grimm, 2013; McReynolds, 2004: 78). With respect to the issue of killing animals, two crucial questions arise from this model of reasoning: which animals share capacities with humans (e.g. having future preferences) that make the end of their lives morally problematic for us? If we can make out relevant capacities in the mentioned sense, death matters morally to animals and killing animals is basically morally wrong. Unsurprisingly, addressing the issue of killing animals from this perspective leads to a rather narrow debate. If all animals are equal on the basis of similar morally relevant capacities, they have to be considered equally. According to this view, treating the phenomenon of 'being in the wrong box' is not just bad luck but an everyday moral disaster.

If we move to a frame of reference other than moral individualism, a second interpretation of the phenomenon of inconsistency seems plausible: addressing the inconsistency as a reasonable, normative plurality. If our relations to animals frame our perspectives on animals (so an animal per se is an illusion) and determine our moral viewpoints, common practices express contingent (but not arbitrary) contexts of justification and social approbation (cf. Crary, 2010; Diamond, 2012). If so, these can be made explicit and can be ethically reflected upon. Therefore, in order to learn about the ethics of killing we shift our attention from what dying and death means to the animal as a natural entity to the contextual preconditions of the notion of animals and the relational practices of killing within certain contexts.

This alternative is inspired by John Dewey's view on ethics, phenomenological insights in the lifeworld as ground and horizon for theory and moral reasoning, and the critique of moral individualism in animal ethics. According to Dewey, an ethical theory finds its origins in three independent and irreducible factors of morals: (1) individual ends (consequentialism); (2) demands of communal life (deontological theories); and (3) social approbation (virtue ethics). Dewey hypothesizes that these three factors have a sound basis and bring about moral conflict because all three play a major role in moral conflict without having a common denominator:

> Posing this point without undertaking to prove it, I shall content myself with presenting the hypothesis that there are at least three independent variables in moral action. Each of these variables has a sound basis, but because each has a different origin and mode of operation, they can be at cross purposes and exercise divergent forces in the formation of judgment. ... [I]t is characteristic of any situation properly called moral that one is ignorant of the end and of good consequences, of the right and just approach, of the direction of virtuous conduct, and that one must search for them (Dewey, 1930: 280).

Whereas the debate on killing animals has often been framed within consequentialist and deontological frameworks with a strong link to moral individualism, we aim at a more contextualized analysis inspired by social practices linked to virtue ethics as a third major aspect in morals. Therefore, the analysis tries to shed light on particular practices of killing and on actors in particular contexts: animal research, slaughtering, and euthanizing pets. Thus, the article aims to broaden the debate on the question of killing animals from a perspective that takes the viewpoint and role of the human lifeworld seriously. By integrating socio-cultural aspects related to particular practices of killing that are embedded in our lifeworld, we try to shift the focus to gain a better understanding of their (complex) normative infrastructure.

6.2 Death: frustration of future preferences

'Dying' can be defined as the 'terminal bodily process of a still living individual'. This process is oftentimes, but not necessarily, accompanied by suffering or pain. In such cases, dying (not death) is a state of immediate experience; death is a state of a body. Thus, death is and can only be experienced by others, but not in the way of knowing what it is like for the other to be dead. It is controversially discussed if animals have a relation in any way to the death of others (fear, grief) or to their own death (angst), and if so, which animals. However, the explanation of the above-mentioned contradictions builds on these concepts of 'death' where the frustration of preferences is vital (cf. Birnbacher, 2006; O'Sullivan, 2011; Regan 2004; Rippe, 2012: 113; Singer, 2011; Wolf, 2012). This view includes animals and the moral respect for their lives only on the basis of their mental capacities. Such an approach can be found in utilitarian theories like Peter Singer's as well as in deontological theories like Tom Regan's. These dominant approaches provide an explanation of ambiguities in human-animal relations in terms of irrationality: If we consider future preferences morally relevant, it does not matter 'where' they are. Human and animal future preferences count equally if they are equal. Therefore, killing – in the sense of frustrating these preferences – matters morally. This line of thought follows the aforementioned extension model. On so-called neutral grounds (Singer, 2011: 90), i.e. outside of any specific practice and normative infrastructure, it is obviously immoral to consider equally equipped beings differently.

The striking similarity of Singer's and Regan's approach is the fact that they both make a morally relevant distinction between animals with the cognitive abilities to have future preferences or an understanding of their own future and those that fall short in this respect (even if they have the abilities on a basic level). Both authors focus on death as the morally relevant phenomenon and frame the debate from this perspective. Consequently, killing is prima facie morally wrong *if* being killed implies the termination of future preferences (Regan, 2004: 324; Singer 2011: 81ff). Cognitive capacities play the role of a selective and exclusive criterion for moral consideration of the lives of animals.

We consider both approaches important in their explanatory power but nevertheless reductionist. Animals only appear within a certain socio-cultural context and are, therefore, not animals *per se* in the sense of given, natural entities (cf. Grimm, 2013). Animals are not abstract entities that can be isolated from our relationship to them in specific contexts. Cognitive abilities are meaningful factors in ethical considerations, but do not cover other lifeworld significances like symbolic over-determination (e.g. of meat consumption in religious rituals), the complex appearance of animals as parasites, pets or pests or the danger posed by animals (cf. Husserl, 1973: 625). It is highly likely that moral claims deriving from reductionist approaches towards animals appear as alien in a given lifeworld. The question that arises is then whether the claim for equal consideration of supposedly equal (human and nonhuman) animals is a universally valid claim that has to be integrated into any socio-historical context or whether its critique gives reason to reconsider and open the debate to other perspectives.

6.2.1 Killing as frustration of future preferences: Singer and Regan

In the seminal book 'Practical ethics' Singer states that the proclamation of the sanctity of life is often reserved for humans only (cf. Singer 2011: 71). Singer further puts forward that we use the term person when we speak of individual *homo sapiens*. He criticizes this position and presents an argument according to Locke: The basis of personhood is the capacity of suffering and of self-awareness (a concept of existing over time) or rationality (cf. Singer 2011: 74, 81ff.). If these capacities can be found in animals, it is rationally sound to ascribe diachronic identity and, therefore, personhood to them. Being a person implies preferences concerned with one's own future. Killing terminates a person's future preferences. Since these preferences are of ethical relevance, killing can only be justified by overriding preferences. Killing is basically wrong because death makes it impossible to follow any future preferences.

Singer consciously pays a high price to protect (human and non-human) persons. His distinction of persons and non-persons implies that death matters morally for the first group only. In cases of conflict, the death of non-persons is morally irrelevant and, therefore, always overridden by interests of persons. Non-persons can even be replaced by others and it is morally irrelevant as long as the net pleasure does not change. Singer presents this idea under the name 'replicability argument' in Practical Ethics (cf. Singer 2011: 106f). However, Singer is quite non-speciesist on who counts as a person: the criterion is not belonging to a certain species but the evidence of the mentioned capacities in whatever species. To support the view that non-human animals also have these capacities, Singer quotes a variety of recent research findings in ethology and Frans de Waal's studies on the capacities of chimpanzees regarding self-awareness and awareness of the intentions of others. Comparable behaviours can be found in pigs and even scrub jays, fish and octopuses (cf. Singer, 2011: 98, 103). It is neither the species nor our image of certain animals that count here. Hence, many other species could be considered persons: 'We think of dogs as more human than pigs, but we have already seen that pigs can plan ahead and grasp whether another pig does or does not know the location of food. Are we turning persons into bacon?' (Singer 2011: 102).

Coming back to our critique above, we have to add here that diverse contexts of killing in different societal contexts could bear significances that oppose such a simplistic and unifold understanding. Referring to the cognitive abilities of animals does not solve conflicts the way they should in Singer's view. If moral problems arise because of significances that cannot be reduced to the capacities of animals, such as regarding the mentioned over-determination of meat consumption, traditional life forms of hunting, etc., the simplistic understanding of killing practices falls short in making the infrastructure of moral conflicts explicit. Moreover, this position leads to counterintuitive claims regarding the well-known, notorious debate on Singer's view of the moral unimportance of being human. To our understanding, being human is not a mere biological fact as Singer seems to hold. It cannot be separated from practices that carry significances in socio-cultural contexts (cf. Diamond, 1991). Thus, being human is much more than a brute fact – it is not the mere membership to the species *homo sapiens*.

Tom Regan's rights approach in animal ethics is usually considered a critique of Singer's position. However, in terms of the moral relevance of death, the two theories show substantial similarities. Like Singer, Regan ties the acknowledgement of fundamental rights to living beings with specific characteristics and capacities. In his theory it is the capacity to meet the subject-of-a-life criterion:

> Individuals are subjects of a life if they are able to perceive and remember; if they have beliefs, desires, and preferences; if they are able to act intentionally in pursuit of their desires or goals; if they are sentient and have an emotional life; if they have a sense for future, including a sense of their own future; if they have a psychophysical identity over time; and if they have an individual experiential welfare that is logically independent of their utility for, and interests of, others' (Regan 2004: 264).

While in Singer's theory the weighing of preferences of all members within the moral community is possible and mandatory, Regan turns his argument against weighing preferences. Subjects of a life should be granted basic, inalienable moral rights. They cannot be outweighed by other's interests or preferences. However, Regan foresees the counter-intuitive implications of his egalitarian abolitionist approach and uses a famous thought experiment – the lifeboat case – to illustrate his solution for conflicts within the moral community:[3] Four human adults and a dog are in a lifeboat which would sink if none of these individuals leaves the boat and if the boat sinks, they will all drown in the open sea. Who should die for the others? Regan's answer is clear: 'Death for the dog, in short, though a harm, is not comparable to the harm that death would be for any humans. ... Our belief that it is the dog who should be killed is justified by appeal to the worse-off principle' (Regan 2004: 324). In everyday life, we are usually not faced with decisions of that kind. However, what we find here is a change from an egalitarian to a hierarchic logic within Regan's approach on the basis of capacities attributed within the group of subjects of a life.

Like in Singer, the consequences of such a position, no matter how stringent it is argued for, are alien to our lifeworld. Of course, trolley cases like the lifeboat case are exceptional and do not lead to universal claims for ordinary situations. Nevertheless, Regan tries to suggest that it would be reasonable (not speciesist) to sacrifice the dog because of the higher cognitive abilities of humans (*ibid.*, 325). Would that provide a justification for a decision in such an unlikely – or in any other – case? And suppose it were the other way round – would we then decide to sacrifice a mentally impaired human to safe a clever dog? We do not believe that this would be a justifiable solution within our lifeworld. Imagine a person claiming that the dog has a richer life compared to the mentally impaired human. It is very unlikely that anyone would accept this argument as a moral justification to sacrifice the human even in a lifeboat scenario. Within our normative infrastructure, one would probably say that this represents just

---

[3] This is just a very brief summary of the lifeboat case that can be looked up in Regan (2004: 351). Regan responds to his critics in the foreword of the 2004 edition of 'The case for animal rights'. A thorough critique of his argument has been presented by Benz-Schwarzburg (2012: 210ff).

a cold calculation (cf. Hursthouse, 2000: 152) outside a common understanding of the significance of being human.

6.2.2 Cora Diamond's critique of Singer and Regan

An important critique of Singer's and Regan's accounts on the morals of killing animals can be formulated against the background of Cora Diamond's arguments. She brings forward the argument that we face severe problems if these frameworks are applied, namely that according to Singer it is perfectly consistent to value humans with impaired cognitive abilities less than persons that do not lack these abilities (cf. Diamond, 2012). Diamond shows that we cannot argue for an appropriate moral practice in human-animal interaction based only on hypostatizing partial principles such as the moral relevance of capacities or biological characteristics. A focal point of her criticism is the well-known problem of the consequences for so called marginal cases. If Singer were right, we would have to describe our immense care for and moral consideration of infants and disabled people as disproportionate, irrational or an exception from the rule. In other words, despite its logical consistency, many of us would not accept the practical consequences of Singer's and also Regan's ideas for moral reasons. For example, justifying the sacrificing of a cognitively impaired human in order to save a dog's life in a lifeboat scenario with respect to the argument that the dog's life is richer does not really settle the issue. The justification 'The dog's life is richer than the human's' is not a sufficient justification in our lifeworld, no matter how logically sound it is. Instead of a 'standpoint of the universe' or the suggested 'neutral ground' (Singer, 2011: 90; Singer and De Lazari-Radek, 2014), Diamond relies on a 'human standpoint' and a consideration of relations and related emotions, providing a basis for moral argumentation (Diamond, 2012). In the following, we will try to elaborate this perspective and apply it to different practices of killing animals.

6.3 Practices of killing: taking the socio-cultural contexts into account

We claim that 'killing' does not simply equal 'using a certain killing instrument and ending the life of an animal'. If we did, the related ethical questions would be to ask how much death matters to the animal and whether the killing method is a good one in terms of avoiding unnecessary pain. We think that our practices of killing in different contexts provide a much richer repertoire to reflect on the issue and it is worthwhile to take account of its complexities. We will try to develop a wider concept of the significance of killing an animal than in moral individualist accounts in order to better understand the moral conflicts. Neutral (biological) grounds are not sufficient to understand moral problems in our lifeworld. In order to make this epistemological standpoint clearer, we will start with a historiographical account by Walter Burkert.

In his book *homo necans* the philologist Walter Burkert asserts that killing animals is always in need of justification or exculpation; at the same time, there are commonly accepted possibilities of justification and exculpation (Burkert, 1997). In particular, killing (and eating) animals as a sacrifice and related rituals are immanent in religious practice and may be the historically oldest part of it. The *homo religiosus* appears as *homo necans* from Athens to Jerusalem to Babylon (cf. Burkert, 1997: 9, 18, 21ff). Historical legends describe scenes in which animals follow their free (or at least god's) will and sometimes even nod in approval to their being sacrificed (cf. Burkert, 1997: 11). These fantastic topoi indicate that there was possibly reluctance but clearly embarrassment connected with killing animals long before thinking about future preferences in the mentioned sense. Moreover, the first known critics of the practice of sacrificing animals are as old as the mentioned legends. Burkert names e.g. Zarathustra, Empedokles and the Pythagoreans (cf. Burkert, 1997: 15). The questions to be asked are: Why not just kill and eat animals? Why did people in the ancient world integrate the killing act into rituals? Are rituals a necessary support to answer the unanswerable phenomenon (cf. Huth, 2011: 209), the inevitable discontent with killing animals?

Obviously, killing (or the dying) of an animal in front of us does not leave the involved person untouched. We find a lot of literature that describes this phenomenon, e.g. in novels (Tolstoy, 1958) and also in philosophical texts (Adorno, 2002). However, the dimension of 'direct concern' hardly occurs in literature on animal ethics. Mostly animal ethicists focus on (rational/reasonable) criteria, supposedly stemming from neutral grounds outside our everyday experiences (e.g. Singer, 2011: 90) for an assessment of killing and killing practices. The experience of being affected or touched by the death or killing appears in Singer and others only as an accessory without normative significance – if it appears at all.

The immediate experience of killing and dying animals has significance in our lifeworld before any reference to moral principles or theories (cf. Lévinas, 1987: 72). It affects us on an existential level. This does not equal knowing what it is like to be a dying pig, dog or bug. The perception – from a human standpoint – of different animals and their dying is always predetermined by social significances and common habits which are neither to be neglected nor sacrosanct and thus alterable. We have to assume that there is a plurality of human-animal relations that is not wholly determined by biological capacities. Derrida prominently focuses on that point when he tries to show that there is not only one kind of human-animal boundary and that the word 'the animal' (*animot*) is an unjustifiable reductionism (cf. Derrida, 2010: 68). Thus, different kinds and also degrees of concern in different relationships are to be taken into consideration. The possibility of being affected by animals' dying, the recognition of animals as vulnerable beings and the patterns of justification of killing in our normative infrastructure predetermine the experience of killing animals.

But the lifeworld and its normative infrastructure is in a constant flux. Current and ongoing debates regarding the killing practices testify the contingency of value-based decisions which represent a justification of killing only *prima facie*.

At the same time we know the problematic fact that the death of animals caused by humans is something often considered unimportant or less important with reference to the animal as 'mere animal'. The proverb 'This is only an animal' is a strategy for reifying living beings to reduce concerns at least on the surface. But according to Adorno, this is to be understood as an act of defiance that enables us to hold that this dying animal is just an animal and 'nothing more' (Adorno, 2002: §68, especially p. 118). Thus, we can make a vital effort to turn these beings into mere bodies, mere objects. To turn this reflection upside down, we can say that this denial of concern is an indirect testimony of this concern.

However, routines, cultural traits, political and legal norms and traditions can make those concerns invisible. Therefore, not only the 'biological animal' but routines, traditions and habits in our dealing with animals have to be reflected upon with regard to their normative infrastructure.

6.3.1 Animal research: killing in context of gaining knowledge

Starting from the idea that killing animals gains differing significance in differing contexts, light will be shed on killing practices and their justification. It will become clear that even in areas where high numbers of animals are killed, the specific killing practices follow normative ideas that value animals in a moral sense. Looking at the field of animal research, it seems as if this has not always been the case in the past. As Bernard Rollin stated (cf. Rollin, 2006: 99), the slogan 'animal use in research is not a moral issue, it is a scientific necessity' was more or less common sense up to the 1980s (cf. *ibid.*). In his statement, Rollin refers to Germany in May 2002, when the protection of animals was elevated to constitutional status. According to article 20a of the German constitution, animals shall be protected. As a consequence, animals are protected from being used and killed in scientific procedures unless an 'absolutely essential necessity' is proved. In response to this move, the biomedical community proclaimed a 'black Friday' (cf. Rollin, 2006: 104).

Looking at the present debates, it can be argued that things have changed and a shift has taken place in our normative infrastructure. Of course, research always followed normative ideas, but animal protection gradually became one of them. This is in part indicated by the success of the 3R approach (replacement, reduction, and refinement), already published in 1959 by Russel and Burch (1959). Since then the 3Rs have been institutionalized step by step with increasing influence in many European countries and in animal protection law. Today a moral compromise can be summarized with regard to moral responsibility in animal research (cf. Sandøe *et al.*, 2008: 117). First, only research should be carried out that aims to deliver vital benefits. Second, the welfare of the involved animals is to be looked after as far as possible. A crucial question remains: What is meant by 'vital benefit' or 'absolutely essential necessity' in this context? This is a matter of the mentioned normative infrastructure insofar as it predetermines what we acknowledge as justification of killing animals.

For instance, the new Austrian Act on the Protection of Animals in Research (TVG, 2012) defines legitimate aims (§ 5 TVG, 2012) according to the EU Directive 2010/63 (EC, 2010). Those purposes have to be plausibly addressed and aimed for within the study designs. For example, as a necessary (not sufficient) precondition for approval by the authorities, the objectives of an experiment in question have to be related to legitimate aims like basic research, translational research, applied research, development and testing of drugs etc. In other words, the Austrian Act does not consider every aim and purpose meaningful enough to justify animal use in research. Quite the opposite, the TVG, 2012 also states purposes that are not significant enough to be (§ 4 TVG, 2012) pursued with experiments. To name a few: purposes that can be achieved without living animals (replacement), experiments that lead to data that have been gained already or do not promise new additional knowledge, experiments for testing weapons, experiments that include great apes, non-human primates, experiments that involve severe pain, suffering or distress that is likely to be long-lasting and cannot be ameliorated (with exceptions) and so forth. Besides these paragraphs that set out the limits of legal purposes, a great number of further criteria are to be applied whenever an experimenter aims at a legal purpose. We find guiding principles, such as the one that experiments have to live up to standards of state of the art research and that animals shall be kept in ways that minimize adverse effects. Further, killing methods and methods of anaesthesia and analgesia, staff requirements etc. are addressed.

Looking at the basic structure of justification in animal research, gaining knowledge obviously provides the reason for the use and killing of animals. But it is, of course, not knowledge per se. At least in the modern understanding according to Francis Bacon, knowledge is associated with power, progress, and prosperity. Therefore, societies allow for the pursuit of knowledge and academic freedom (within limits) even if other social, political, pragmatic and moral aims, like the protection of animals, are overridden. On these grounds, this line of justification is only valid as long as the promises of animal research are held.

From this point of view it becomes plausible to ask whether the intense debate about using and killing animals in research is not only fuelled by the fact that animals suffer and are killed. Eroding trust and a decrease in the perceived importance and relevance of scientific knowledge play a major role. Thus, an approach to the ethics of animal use in research should involve a reflection on the kind of knowledge we want and need and, most importantly, what sort of knowledge we can reasonably gain in animal research as well as what promises can be reasonably held. This issue becomes increasingly pressing since the empirical argument that animal research reaches intended goals has been questioned in the recent past (cf. Knight, 2011; Lindl *et al.*, 2006).[4]

Looking at the debate from this point of view, we find that the question of using animals in scientific experimentation leads to questions concerning the relevance of certain kinds of knowledge and the general confidence in science: if scientists do not deliver relevant knowledge, the mentioned justification for using animals decreases. Taking this into account, it is plausible that the actual debate on killing and using animals in research is not only fuelled by the question of moral status, but also by the crisis of confidence in scientific institutions. In short, the question of using and killing animals in research leaves us with the question 'What is the relevant knowledge we are willing to pay for with animal suffering and lives?' and 'What sort of knowledge are we willing to lose in order to save animal lives?' These issues cannot be settled by inquiries into the question of the moral relevance of future preferences as outlined above. Quite the opposite, the above-mentioned approach (moral individualism) demonstrates a position in moral theory that makes moral conflict impossible. The moral problem would not even emerge if the question of animal research were reduced to only one source or principle. Using and killing animals would be wrong and would have to be stopped straight away. Using Dewey's argument in 'Three independent factors of morals' (Dewey, 1930: 279-288), this can be illustrated with the following quote: 'That is the necessary logical conclusion if moral action has only one source, if it ranges only within a single category. Obviously in this case the only force which can oppose the moral is the immoral' (Dewey, 1930: 280). But why are we still asking and reflecting on the issue, if the situation is that clear? The short answer is that it is not that clear at all.

Since the aim of science is to gain knowledge that sooner or later could contribute to a prospering society, first, the question of what kind of knowledge or promised benefit can justify suffering and killing of animals is directly linked to the question 'What is a prospering society?' and 'How can science contribute through animals research?' Science and scientists cannot answer this question on their own.

As a second step, we want to look at the actors involved in animal research. We want to highlight that the individuals involved in the research process are usually not indifferent regarding their work and the animals used. They do their work within

---

[4] As Diamond (1991) has shown, the patterns of justification would be extremely different in case of experiments in humans. The different legal frames regarding the admissibility of human experimentation compared to animals give evidence that the moral structure we are living in distinguishes between humans and animals not according to rational capabilities but according to the different relation we have to human beings than to non-human beings.

the outlined context of justification and are usually convinced that animal research is necessary and justified by overriding objectives (cf. Rollin, 2006: 107). Otherwise, they would probably experience more difficulties with their being affected by animals' dying. Implicitly, this is to say that experimenters accept that a moral problem is given and are affected by harming and killing animals. A particularly striking example is the following quote from a laboratory staff member: 'They did not like having rats in clear cage 'because the animals could look at you' (Linda Birke in Acampora, 2006: 100). In other words, they share the experience that the practices of letting animals suffer and of killing animals are morally and (thus) also psychologically problematic. If this were right, scientists would accept the moral problem but deem their practice justified *because* they are convinced that they have justifying reasons to do so, even if they see that there is a moral conflict.

After addressing this rather fundamental topic, three relevant practices of killing in the context of animal research shall be sketched in order to further illuminate the normative infrastructure of that field.

1. *Killing indicated by humane end-points.* In article 13 of the EU Directive 2010/63 (EC, 2010) humane end-points are addressed: 'Death as the end-point of a procedure shall be avoided as far as possible and replaced by early and humane end-points'. Accordingly, death, in the sense of dying without the help of humans, shall be explicitly avoided. Instead, killing seems to be the favourable method. The reason is quite obvious: dying can imply severe suffering and pain for the animals. In order to minimize pain and suffering, humane end-points shall be defined at which animals are 'taken out of the experiment'. For instance, this could be a severe loss in body weight of more than 20% or indication of severe pain, etc. Looking at these humane end-points more closely makes quite clear that the life of animals is not considered as important as the avoidance of pain and suffering in this context. Humane end-points are considered a refinement method in accordance with the 3R principle. The normative idea about animals is obviously not the 'living animal' but the 'non-suffering animal'.

2. *Killing at the end of an experiment.* In a similar vein, killing animals at the end of an experiment demonstrates that the principle to avoid adverse effects on the wellbeing of the animal overrides the principle of saving animal lives. The question of killing is not seen as a moral problem in itself but as a solution to the problem of causing pain, distress, etc. Therefore, the preamble of the EU Directive 2010/63 (cf. EC, 2010) calls on professionals to kill animals in the following way:

> The use of inappropriate methods for killing an animal can cause significant pain, distress and suffering to the animal. The level of competence of the person carrying out this operation is equally important. Animals should therefore be killed only by a competent person using a method that is appropriate to the species.

In this case, the focus also lies on the animal's ability to experience pain and distress and not on killing and protecting their lives.

3. *Killing in the category non recovery.* The same applies to experiments in the category non recovery. Those experiments are defined as procedures which are performed entirely under general anaesthesia from which the animal shall not regain consciousness. Since these procedures promise no pain and distress, they are usually considered 'mild' in the severity classification, meaning that procedures that cause moderate pain, suffering, or distress are classified as more problematic than the killing of an animal in our current normative infrastructure.

Adding the normative similarities of the mentioned killing practices together, one can see that the life of animals is not as highly valued as the avoidance of pain, suffering, and distress in this context. This also follows a rather brutal logic in the field of animal research: only adverse effects that are related to the scientific purpose can be justified. If the data are gained and the procedure is over, there is no justification other than a new scientific procedure that can save the animals' life. On the one hand, research interests keep the animals alive as long and at a high standard as possible within the context of animal research. On the other hand, as soon as there is no research interest anymore, there is no reason to save the animals' lives.

In short, the normative idea applied to lab animals is not a 'living being' but a 'suffering being'. This was shown by the normative justification within the killing practices that usually find their reason in the animals' interest to suffer as little as possible. This logic is, of course, only convincing within the context of animal research and not so plausible when looking at it from the 'outside'. This leads us back to the more fundamental debate on the basic justification of animal research and whether the (legal) framework is set right. The logic within the existing framework is rightly described as 'using animals as research instruments and trying to minimize pain, suffering and distress'. The suffering animal is the normative benchmark. This logic often leads to frustration and massive critique for people not working in the field since the normative idea of animals is not only a 'suffering being' but also a 'living being'. From this outside perspective, experimenters do not live up to normative expectations. From an inside perspective, researchers do everything they can in favour of the animals. Unsurprisingly, this must lead to a conflict and shows the immanent complexity and multi-perspectivity of normative infrastructures.

Using the idea of 'Three independent factors of morals' (Dewey, 1930: 279-288), we see that we are dealing with a much wider moral problem than the moral relevance of death. We find (a) individual ends, such as the aim of an academic scientist to gain knowledge or a PhD in a field where academic freedom is granted, or a private lab owner who wants or has to make money. Such aims are not at all undisputed. On the contrary, there are only few justifying reasons left that are likely to be accepted as good reasons for killing animals in the research process. Secondly, we see that in Dewey's heuristic (b) demands of communal life are at stake. If the lives of animals have a moral standing, academic freedom as a vital principle of a 'knowledge-based society' is not only restrained by economic limitations or other principles that contribute to our social welfare but also by concerns about animals. The corresponding question is how far we are willing to adjust or alter a supposedly highly successful strategy or rethink the entire strategy of a knowledge-based society. Social approbation, the third factor in Dewey's trilogy, leads to further considerations: Scientists who were unquestioned in their contributions to society are confronted with critique of being inconsiderate moral outlaws (cf. 'Der Tagesspiegel', a German newspaper, refers to the debate about the German scientist Andreas Kreiter being attacked because of his research on monkeys and even the necessity of police protection for his family[5]). The changes in the human-animal relationship question roles and responsibilities in society. This has a vital impact on people working in that field and on social institutions.

6.3.2 Slaughtering: killing for taste and tradition

Generally speaking, a vast majority of Europeans are meat eaters. At the same time it is almost unimaginable that there is a slaughterhouse in any western civilization downtown. This describes the structural invisibility of killing animals (cf. O'Sullivan, 2011). Eating meat is common, but killing for food does not belong to everyday life; it is out of (the moral) view. Videos and pictures of industrial killing practice might lead to a change in our point of view and maybe consequently to a change in eating habits, but usually the images that bring invisible killing to light are not part of our everyday consciousness (cf. Joy, 2010). They come up in shock documentaries that appear as exceptions to our everyday routines. The eating habits of a majority are touched by these pictures only to a little degree. (Similarly, watching the football World Cup might be impressive, but does not result in an immediate change of habits with regard to football.) We will again take up the stance that it is not clear that eating meat per se is bluntly morally wrong and talk about what our practices look like and the question how we could and should shape and change them.

Nevertheless, as a reaction to the striking number of killings and the industrial methods, steps have been made to minimize the animals' suffering (under conditions of the notion/construction of animals being able to suffer with moral relevance). Death is basically caused by the withdrawal of blood after stunning. In Germany, for instance, stunning has been required by law since 1986 in order to prevent animal pain (cf. Troeger, 1997). As a consequence, differentiations of legitimate and illegitimate (thus cruel) killing practices have been made – but not only according to

---

[5] Available at: http://tinyurl.com/odpq8gp.

legislation but also according to common understandings (cf. Sebastian, 2013: 105). Moreover, industrial slaughtering under conditions of efficiency is also described in terms of traumatization of slaughterhouse staff. This is only understandable against the background of our reflections about the concern by killing/dying of animals in the beginning of Section 6.3. Otherwise we would consider it squeamish or pathological that these employees report on psychological burdens. A particularly drastic description is given in Drillard's article 'A slaughterhouse nightmare' (Drillard, 2008). She quotes a person working at the frontline of slaughtering: 'the worst thing, worse than the physical danger, is the emotional toll ... Pigs down on the kill floor have come up and nuzzled me like a puppy. Two minutes later I had to kill them – beat them to death with a pipe. I can't care' (Drillard, 2008: 391). Here it is obvious that hardening can – if any – succeed only in part. On the one hand, 'I can't care' expresses the structural impossibility of taking time to reflect on the event – not at least because of the given work task to kill animals. On the other hand, it manifests a radical helplessness because of the concern and impossibility of answering 'adequately' (e.g. within rituals) to the dying of a great number, caused by one's own hands.[6] Such reports make resignation visible. Thus, it is not surprising at all that one possible consequence of such a confrontation with the killing of animals is posttraumatic stress disorders or burnout symptoms (cf. Huth, 2014: 68f).

At the same time, there are exceptions in case of rituals and religious practices of killing regarding the mentioned stunning practice. Religious traditions can conflict quite easily with a policy of (supposedly) painless killing and with basic questions of the human-animal relationship. Against the background of pathocentric approaches, the ritual of killing conscious animals is hotly debated; the *schechita* is under discussion. Rituals that are linked to harming and killing animals determine a field in which spiritual practice (practical religious freedom) partly clashes with classic animal welfare concerns reflected in dominant theories of animal ethics and in animal protection acts. Once more, we see the multidimensionality of a pluralistic normative infrastructure.

The background of killing for meat is a heritage that stems from the history Walter Burkert has described in great detail. Meat is not just nutrition. It still carries a surplus of meaning. It has meanings of luxury, wealth and celebration. In terms of nutrition, we can easily find alternatives to meat, but not for its symbolic significance. Therefore, higher prizes or a meat shortage would have a number of social and cultural consequences which have nothing to do with malnourishment. The over-determination of meat, preserved in increasingly questioned traditions, legitimates the practices of industrial slaughtering at least to a certain extent. From the ethical point of view that refers to the moral problem of killing in terms of future preferences,

---

[6] The societal unacceptability of factory farming and relative killing practice might derive from an implicit acknowledgement of the pathos for the involved by killing animals in that way.

it is clear that we should stop producing meat. In our view, in order to come closer to that goal, the issue should not be reduced to the question of killing.

This brings us once more back to Dewey's trilogy: considering individual ends, we see that most Western societies do eat meat on a large scale. Since they could also be on a meatless diet, the question arises if taste can justify, for instance, the slaughtering of roughly 60 million pigs per year in Germany. Are taste and tradition sufficient to kill animals? We think that the demands of communal life – as the second dimension – would give a clear answer: killing for food is legitimate (and legal), even if we have alternatives – although this is not entirely unquestioned. There is no clear contradiction between reasonable argumentation and unreasonable tradition as considered in the positions of moral individualism. The third dimension – social approbation – seems to decrease step by step. Eating meat or not eating meat increasingly becomes a moral statement in the public sphere. However, we are in the middle, not at the end of a societal debate (and we assume that an end of this debate is not foreseeable). It is not very likely that the public debate will be settled by arguing against the evils of terminating future preferences. The question is how to change existing traditions as well as why and whether new habits can be developed that compensate for the symbolic dimensions of eating meat. *Prima facie,* we can say that there is discomfort regarding some practices, especially in industrial slaughtering. This discomfort may not stem from an absolute (meant literally as ab-solute – out of context) principle or imperative, but from the gap between the direct concern for dying animals and the radical reification in some slaughter practices. The ethical squinting for absolute principles and clear-cut solutions creates potential and actual frustration that counteracts basically reasonable arguments. They appear alien with little argumentative force in our lifeworld. At the moment, a general prohibition of meat consumption would – however well-argued, perfectly reasonable and compelling – lead to heavy opposition and be considered totalitarian for good reason.

At the same time, we should recall the discomfort people feel when facing industrial meat production and hearing about the immense numbers of animals slaughtered because of taste and eating habits. Industrial meat production would not face the social approbation it faces today, if at least a majority within our lifeworld could experience the killing of animals for meat. Then, our concerns could lead to negotiations (discourse) regarding the moral significance. Probably, this would influence our opinions and habits. The hypostatization of certain characteristics or capabilities like the ability to suffer or having future preferences might be justified in certain arguments, but cannot live up to the complexity and socio-historical contingency of lifeworld significances and common habitual conduct. Maybe one day any kind of killing animals for nutrition will not be socially acceptable any more.

6.3.3 Euthanasia of companion animals: killing to be good

What is the animal in a close human-animal relationship (and not: per se)? It can clearly not be reduced to a biological organism with future preferences. A survey in the USA demonstrated quite clearly that animals are family members: 'Some feel closest to their pet. Indeed, in a national survey 57% of respondents, if stranded on a desert island with only one companion, would choose their family pet'. (Walsh, 2009: 481) Walsh describes a human-animal relationship that differs significantly from those in the context of animal testing and meat production. Companion animals and their owners live in close contact with strong emotional bonds. Not only cats and dogs but also 'fids' (feathered kids) amongst others are supposed to have important psychological and socio-emotional impacts like increasing children's empathy and responsibility. They sometimes even replace children with many similar implications (cf. Walsh, 2009: 481f).

Thus, it is not surprising that a pet's death often causes long mourning phases. The intensity of emotional bonds makes it hard to take farewell of the animal companion or family member. Walsh notes that 85% of pet keepers show significant symptoms of grief. Moreover, in a great number of cases it is the first loss, the first confrontation with death that children experience in their lives. The author adds that, nevertheless, these processes are frequently underestimated. People mourning for their pets are often still considered ridiculous and, therefore, hide their emotions (cf. Walsh, 2009: 487).

Taking into account that humans (children) do not learn to interact with animals on farms anymore but in private households with animals as family members, the change of perspective on animals and conflicting viewpoints regarding killing can be better understood. Only a few decades ago it was not a problem to kill a seriously ill dog or cat, even by one's own hands. The old and blind watchdog was 'simply' shot. Cats were not sterilized, but the offspring were killed by drowning, shooting, etc. if there was no use for more cats.

This has fundamentally changed in human-pet relationships: 'increasingly, those who are strongly attached to their pet are electing costly, extensive medical treatments now available, unless the animal's suffering or the caregiving burden becomes too great' (Walsh, 2009: 489f). Only if the animal's suffering becomes severe and cannot be eased, does it serve as a justifying reason to kill an animal (cf. McMahan, 2002: 201). The family cares, the vet kills (eases the suffering). 'Studies find that for most, euthanasia is more beneficial for both the animal and human companions than waiting for a suffering pet to die 'naturally'' (Walsh, 2009: 490).

Euthanizing pets is one of the most problematic aspects of a vet's responsibilities. A recent study among Austrian vets indicates that nearly 13% get used to euthanizing animals. An equal share (little more than 13%) never gets used to it. Another noteworthy element of this poll is the fact that more than 40% approved that respectful treatment of euthanized animals is an inherent part of good practice (compared to only 1.47% who denied that it is) (Springer *et al.*, 2013). This leads to the assumption that the dead body of a pet is a matter of piety, not of irrelevance and of course not of hunger.

Looking at the killing practice in the field of pet animals, it becomes rather clear that killing and death is something rather different in this context when compared to animal research or slaughtering. In its most unquestioned form, euthanasia is a practice to end a painful existence. Shifting light on the involved parties, it becomes clear that the owner can demonstrate a praiseworthy moral attitude in this situation – not just because the death of a pet animal brings about the loss of a family member that has to be dealt with. The owner's consent or expressed will that the animal shall be killed by a veterinarian can be seen as an advocatory practice in favour of the animal. The following case of a severely ill dog in pain illustrates the justification of killing: The vet in his or her role of a professional diagnoses an illness that gives medical indication to kill the animal. The owner is now in the dilemma of having to decide either that the animal should be killed in order to prevent the animal from suffering or that the animal should be allowed to live in pain. As a responsible owner, he or she will take the perspective of the dog and ask what the dog's will would be. In severe cases, this leads to the decision to euthanize the animal in order to end its suffering. Following this kind of logic, the owner and the vet are 'helping' the animal to meet its presumed will. Sentences like 'It is better for her to die' make this fact explicit. The best justification is given if the killing of the animal is in its own interest. Obviously, this tells us more about our perception of animals then about the animal itself. However, the point is that in such situations pet owners demonstrate a morally highly valued attitude and virtue: mercy.

The merciful person does not use his or her power to use the powerless, but uses power to help and protect them. What we find here is a strong motive in our moral life. Milan Kundera describes it in the following passage:

> True human goodness can manifest itself, in all its purity and liberty, only in regard to those who have no power. The true moral test of humanity (the most radical, situated on a level so profound that it escapes our notice) lies in its relations to those who are at its mercy: the animals. And it is here that exists the fundamental failing of man, so fundamental that all others follow from it (Kundera, 1984: 328f, transl. Rowlands, 2012: 31).

Without going into great detail, the case of euthanizing pets is a perfect example where pet owners can demonstrate their humanity in a merciful killing practice. Whereas other cases of killing are perceived as problematic, euthanasia of pets serves as a normative ideal for good killing and shows a normative infrastructure of justified killing. The owner, ruling over life and death, decides according to the presumed will of the animal. There is nothing wrong with this practice and the owner demonstrates humanity at its best, since it is the will of the animal (and not the owner's). This form of killing is usually not questioned at all and serves as a role model for other killing practices. Often it is even seen as a moral duty to end an animal's life and other killing practices are criticized against this background: for instance, the illustrated practices of slaughtering and killing at the end of an experiment can definitely not be described as merciful killing according to an animal's will. To carry it one step further, culling of healthy animals (zoonosis) to protect the meat market is an example that illustrates a killing practice that is just the opposite of killing in a merciful attitude: Powerful professionals are putting down animals in the thousands for other than the animal's will. Using power against the powerless is to be described as an unmerciful action. Also slaughtering and killing in the context of animal research can be structured along this line of thought. Whenever the killing practice is beyond the suspected will of the animal, it becomes a problem. Whenever it comes close to euthanasia of pets, the moral problem decreases and the moral agent makes himself or herself a praiseworthy person.

Coming back to Dewey's trilogy, keeping pets and treating them well are signs of a caring, considerate, and thus virtuous character. Social acceptance can be reached even through killing. In this case, killing is a form of doing good and not just ending preferences. Moreover, it can be considered as the fulfilment of an animal's suspected preferences and a normative ideal.

6.4 Conclusions

Pragmatism and phenomenology correlate especially in one important point: they proceed from embedded practices in socio-cultural contexts. Therefore, differing practices of killing carry differing significances and determine our perception and notion of humans and animals and their moral relationships. We called this the normative infrastructure of killing practices. Thus, it is necessary for our approach to take habits and cultural traits into account. Human behaviour is not guided by decisions according to principles alone. Different contexts can lead to different moral demands and inconsistent practices for good reason.

From this point of view, an alternative would be the – on the first glance maybe discomforting – account of relational animal ethics. The animal per se as a natural entity is fiction, although science could inform us in a very sophisticated way about animal needs and capacities (Grimm, 2013). But at the same time the human standpoint is a culturally determined perspective which can be influenced but not fully determined or constructed by arguments, no matter how rational they are (cf. Huth, 2013: 123-127). We do not support relativism or 'anything goes', but try to make plausible that supposedly universal claims from neutral grounds themselves stem from socio-cultural contexts. They are a product of historical settings in which, for instance, the suffering is considered the worst for animals and cognitive abilities (as shaping the quality of suffering) build the basis for moral consideration. This viewpoint does not cover our relation to animals. A lack of consideration of this point leads to counter-intuitive appeals and to conduct and conditions that are currently impossible for a major part of our society. Such claims could even appear as a form of cultural imperialism.

Taking relationality seriously does not mean to immunize traditions and habits against critique. Moral conflict also arises within traditions and habits. Thus, making the normative infrastructure of practices of killing animals visible is a prerequisite for an informed debate that can lead to democratic decisions a society can also bear.

References

Acampora, R., 2006. Corporeal compassion. Animal ethics and philosophy of body. University of Pittsburgh Press, Pittsburgh, PA, USA, 224 pp.

Adorno, T.W., 2002. Minima Moralia. Reflexionen aus dem beschädigten Leben. Suhrkamp Verlag, Frankfurt am Main, Germany, 302 pp.

Aerts, S., Bonnen, R., Bruggeman, V., Tavernier, J. and Decuypere, E., 2009. Culling of day-old chicks: opening the debates of moria? In: Millar, K., West, P.H. and Nerlich, B. (eds.) Ethical futures: bioscience and food horizons. Wageningen Academic Publishers, Wageningen, the Netherlands, pp. 117-122.

Benz-Schwarzburg, J., 2012. Verwandte im Geiste – Fremde im Recht: Sozio-kognitive Fähigkeiten bei Tieren und ihre Relevanz für Tierethik und Tierschutz. Harald Fischer Verlag, Erlangen, Germany, 481 pp.

Birnbacher, D., 2006. Bioethik. Zwischen Natur und Interesse. Suhrkamp Verlag, Berlin, Germany.

Bulliet, R., 2005. Hunters, herders and hamburgers: the past and the future of human-animal relationship. Columbia University Press, New York, NY, USA, 264 pp.

Burkert, W., 1997. Homo necans. Walter de Gruyter GmbH, Berlin, Germany, 378 pp.

Crary, A., 2010. Minding what already matters: a critique of moral individualism. Philosophical Topics 38: 17-49.

Derrida, J., 2010. Das Tier, das ich also bin. Passagen-Verlag, Vienna., Austria.

Dewey, J., 1930. Three independent factors in morals. In: Boydston, J.A. (ed.) The later works, volume 5, 1929-1939: essays, the sources of a science of education, individualism, old and new, and construction and criticism. Southern Illinois University Press, Carbondale, IL, USA, pp. 279-288.

Dewey, J., 1922. The middle works of John Dewey, volume 14, 1899-1924: human nature and conduct. Collected works of John Dewey. Southern Illinois University Press, Carbondale, IL, USA.

Diamond, C., 1991. The importance of being human. In: Cockburn, D. (ed.) Human beings. Cambridge University Press, Cambridge, UK, pp. 35-62.

Diamond, C., 2012. Menschen, Tiere und Begriffe. Aufsätze zur Moralphilosophie. Suhrkamp Verlag, Berlin, Germany.

Drillard, J., 2008. A slaughterhouse nightmare. Psychological harm suffered by slaughterhouse employees and the possibility to redress through legal reform. Georgetown Journal of Poverty Law and Policy 15: 391.

European Commission (EC), 2010. Directive 2010/63/EU of the European Parliament and of the Council of 22 September 2010 on the protection of animals used for scientific purposes. Official Journal of the European Union L 276: 33-79.

Grimm, H., 2013. Das Tier an sich? Auf der Suche nach dem Menschen in der Tierethik. In: Rippe, K.P. and Thurnherr, U. (eds.) Tierisch menschlich. Harald Fischer Verlag, Erlangen, Deutschland, pp. 51-96.

Herzog, H., 2011. Some we love, some we hate, some we eat: why it's so hard to think straight about animals. Harper Perennial, New York, NY, USA.

Husserl, E., 1973. Zur Phänomenologie der Intersubjektivität: Texte aus dem Nachlaß Volume 3. Springer, Dordrecht, the Netherlands.

Huth, M., 2011. Den Anderen behandeln und betreuen. Phänomenologische Ansätze zu Grundfragen der Medizin. Verlag Karl Alber, Freiburg, Germany.

Huth, M., 2013. Negative Integrität. Das Konzept der Leiblichkeit in der Ethik der Mensch-Tier-Beziehung. TIERethik 1: 108-128.

Huth, M., 2014. Ihr Tod geht uns an. Eine Phänomenologie des Sterbens von Tieren. In: Ullrich, J. and Ulrich, A. (eds.) Tierstudien 05/2014. Neofelis Verlag, Berlin, Germany, pp. 59-71.

Joy, M., 2010. Why we love dogs, eat pigs, and wear cows: an introduction to carnism. Conari Press, Newburyportm MA, USA.

Knight, A., 2011. The costs and benefits of animal experiments. Palgrave MacMillan, New York, NY, USA.

Kunzmann, P., 2013. Sich wandelnde Verhältnisse zum Tier – Wandel im Tierschutz. TIERethik 1: 55-77.

Lindl, T., Völkel, M. and Kolar, R., 2006. Animal experiments in biomedical research. An evaluation of the clinical relevance of approved animal experimental projects: no evident implementation in human medicine within more than 10 years. ALTEX 23: 111.

McMahan, J., 2002. The ethics of killing. Problems at the margins of life. Oxford University Press, New York, NY, USA.

McReynolds, P., 2004. Overlapping horizons of meaning. A Deweyan approach to the moral standing of nonhuman animals. In: McKenna, E. and Light, A. (eds.) Animal pragmatism. rethinking human-nonhuman relationships. Indiana University Press, Bloomington, IN, USA, pp. 63-85.

O'Sullivan, S., 2011. Animals, equality and democracy. Palgrave MacMillan, New York, NY, USA.

Regan, T., 2004. The case for animal rights. University of California Press, Berkeley, LA, USA, 474 pp.

Rippe, K.P., 2012. Ein Lebensschutz für Tiere? In: Michel, M., Kühne, D. and Hänni, J. (eds.) Animal law – Tier und Recht. Developments and perspectives in the 21st Century. Entwicklungen und Perspektiven im 21. Jahrhundert. Dike Verlag, Zürich, Switzerland, pp.87-116.

Rollin, B., 2006. Science and ethics. Cambridge University Press, New York, NY, USA.

Rowlands, M., 2012. Virtue ethics and animals. In: Protopapadakis, E. (ed.) Animal ethics: past and present perspectives. Logos Verlag, Berlin, Germany, pp. 29-38.

Russel, W.M.S. and Burch, R.L., 1959. The principles of humane experimental technique. Methuen Publishing, London, UK.

Sandøe, P., Christiansen, S.B., Hansen, A.K. and Olsson, A., 2008. The use of animals in experiments. In: Sandøe, P. and Christiansen, S.B. (eds.) The ethics of animal use. Blackwell Publishing, Oxford, UK, pp. 103-117.

Sebastian, M., 2013. Tierliebe im Schlachthof? Das Interesse am Wohl der Tiere als Verarbeitungsstrategie von Gewalt im Schlachthof. Tierstudien 3: 102-113.

Singer, P., 2011. Practical ethics. Cambridge University Press, New York, NY, USA.

Singer, P. and De Lazari-Radek, K., 2014. The standpoint of the universe. Sidgwick and contemporary ethics. Oxford University Press, Oxford, UK, 432 pp.

Springer, S., Hartnack, S., Weich, K., Moens, Y. and Grimm, H., 2013. Euthanasie in der Kleintierpraxis – Ergebnisse der TierärztInnenbefragung. In: Baumgartner, J. (ed.) Tierschutz. Sektion Tierhaltung und Tierschutz der Österreichischen Gesellschaft der Tierärzte, Vienna, Austria, pp. 21-28. Available at: http://tinyurl.com/ookq46y.

Tolstoy, L., 1958. Krieg und Frieden. Bertelsmann Verlag, Gütersloh, Germany.

Troeger, K., 1997. Schlachten von Tieren. In: Sambraus, H.H. and Steiger, A. (eds.) Das Buch vom Tierschutz. Enke Verlag, Stuttgart, Germany.

TVG, 2012. Bundesgesetz über Versuche an lebenden Tieren. Bundesgesetzblatt I Nr. 114/2012.

Walsh, F., 2009. Human-animal bonds II: the role of pets in family systems and family therapy. Family Process 48: 481-499.

Wolf, U., 2012. Ethik der Mensch-Tier-Beziehung. Verlag Vittorio Klostermann, Frankfurt am Main, Germany.

7. Even a cow would be killed …: about the difference between killing (some) animals and (some) humans

F.R. Stafleu

Ethics Institute, Utrecht University, Janskerkhof 13, 3512 BL Utrecht, the Netherlands; f.r.stafleu@uu.nl

Abstract

This essay raises the question why there is a difference between the way we treat animals and humans, when it comes to killing. The question is analysed with the help of two special cases. On the one hand, a non-autonomous patient whose suffering is immense and hopeless. On the other hand, an old dog that equally suffers badly. The differences and similarities are analysed and discussed from the perspective of ethical theory. The discussion includes an analysis of the taboo on killing humans and the possible biological explanation for this phenomenon. It is argued that overriding this taboo causes existential moral doubts. This burden can serve as a moral justification for operating (even) more cautiously in case of the human patient. The conclusion has an impact on both our dealings with animals and humans.

Keywords: taboo, patient, suffering, autonomy

7.1 Introduction

In the neurology department of my wife (who is a neurologist) an old farmer was taken in. He was terminally ill, but not yet in his last days. He suffered, but not unacceptably and he was taken good care of. Nonetheless, when my wife visited him he said: 'Dear doctor, even a cow would be mercifully killed in my situation, but a fellow human you let die slowly'.

Why we treat humans and animals as it comes to killing totally different in a situation like above described, has intrigued me (a veterinarian who is specialized in ethics) for many years. Writing this chapter has been an opportunity for a first exploration of the subject which materialized in the shape of an essay. The difference between killing humans and animals is an enormous subject with many different angles: slaughter, hunting, war, human-animal relationships in general, the value of human and animal life, etc. So I had to focus on *some* animals and *some* humans in the particular situation of the end stage of life like in the example of the farmer and the cow above. The backbone of this essay is the comparison between two cases.

*Franck L.B. Meijboom and Elsbeth N. Stassen (ed.) **The end of animal life: a start for ethical debate***
DOI 10.3920/978-90-8686-808-7_7, © Wageningen Academic Publishers 2016

7.1.1 Case 1. An Alzheimer patient (89 years old)

She lives in a nursery and the Alzheimer is in a progressed stage, but the disease is not directly life threatening. Her cognitive abilities have gone. She suffers from attacks of anxiety and seems to be lonely. She is often found at the (closed) door because she wants to get out. Lately she was diagnosed with colon cancer. She suffers from arthrosis, and is not responsive to painkillers. Good care does not substantially alter this very unhappy situation. Caretakers, doctors and family are convinced that this patient suffers seriously.

7.1.2 Case 2. An old dog

This dog has breast cancer and arthrosis. The dog stays in her basket, does not want to go out any more and reacts slowly when the owners try to make contact. Walking seems painful and the pain seems hardly to be relieved by painkillers. The tumour has grown considerably, but is not directly life threatening for the dog. An operation is not possible, the dog is too old. The owners and veterinarian are convinced that the dog is suffering and that there is no hope for recovery.

In the case of suffering without hope, veterinarians feel morally obliged to kill an animal (KNMvD, 2010). In general the discussion is about deciding when to kill the dog. The right moment is when the dog begins to suffer or suffers too much. When that moment has arrived, it is in general morally accepted or even considered to be a moral obligation to kill the dog. The dog is not killed earlier because most owners and veterinarians respect the life of the dog, but this respect is overridden by the principle that aims to avoid unnecessary suffering.

In the case of the human patient, family and doctors go to great lengths to ameliorate the suffering. When that fails they hope for a 'merciful death', caused for instance by a pneumonia. The wish for 'a merciful death' indicates that they think that the end of suffering is preferable above surviving. So they think it is in the best interest of the patient that she dies. But still they will not actively choose to end the suffering by killing her. There seems to be a strong moral inhibition concerning killing humans that (in our case) trumps the avoidance of the suffering of that patient.

Why this difference? At first sight the cases of the dog and the patient are comparable: both suffer without hope and the cognitive abilities of both are on a comparable level. So why do we feel morally obliged to kill the dog, but not the patient? As mentioned above, this question puzzles me and motivated me to write this essay.

When I pose this question in a discussion, I often get two kinds of answers. Doctors and medical ethicists (in most cases) think it is a strange or even an evil question. For them it is self-evident that there are relevant differences between humans and animals. Often I am not taken seriously and the discussion comes to a halt. Although I do recall that once a doctor (also a medical ethicist) mentioned the 'holiness of human life', but what this entailed remained vague. The second reaction often comes from nurses and caretakers. They say 'you are right, there ought not to be a difference in these cases, indeed dogs are better off than people!' When I ask: 'Would you then be willing to kill them yourself when it were legally allowed?' they often hesitate. So, in these discussions, there seems to be two conflicting moral intuitions at stake: 'Human life is special, so do not kill' versus 'killing this kind of patients is the right thing to do'.

In the above described circumstances the killing of the dog seems morally undisputed. The (not) killing of the patient causes hesitation, at least by me and the caretakers/nurses. Therefore I will mainly focus on the question why we do not kill the patient, by searching morally relevant differences and similarities between the human and animal patient.

My hypothesis is that there is a taboo on killing humans and not a taboo on killing dogs. I call it a 'taboo' to emphasize the almost absolute nature of it, mirrored by the reference to the 'holiness of life' by the medical ethicist and the hesitation of the nurses and caretakers. In this essay I look for the biological and moral basis of this taboo in order to answer my basic question: is it morally right that we do not kill the human patient?

7.2 Taboo on killing humans, a biological basis

In the former section, I used the word taboo in the sense of untouchable, sacred. This is the original Polynesian meaning of the word. It could also mean something that is prohibited by social custom (Encyclopaedia Britannica, undated). When I talk about a taboo on killing, I aim to emphasise that it is a near absolute or at least very strong (social) rule on not killing people.

What are the roots of this taboo? As a veterinarian I always start by looking for a biological explanation. Not because I believe that such an explanation can be directly translated in moral norms, but because I think that biological drives are the ontological basis of morality. Frans de Waal, an ethologist, has given ample arguments in his publications that the origin of our morality must be explained by the social nature of man (De Waal, 2008). The social group of the chimpanzee (evolutionarily closest to us) is a complex structure in which individual and social interests are in a delicate equilibrium. This equilibrium is maintained by a fairly complex set of social rules and biological drives. Frans de Waal states that this complex social structure is a kind of 'animal morality', which is the biological basis of human morality. Human ethics however is more evolved, basically in the sense that humans can reflect on social rules and biological drives (De Waal, 2008). In other words humans can distinguish between how it *is* biologically and how it *ought to be* from an ethical point of view. This unique human capacity to reflect makes the difference between 'animal morality' and ethics: ethics is systemic reflection on morality. If the morality on which we ethically reflect has a (partly) biological basis, it is important to know what this basis is, for two reasons. Firstly, the original biological function might help us to identify ethically relevant arguments. Knowing the origin of moral intuitions may give insights that lead to a better and richer ethical reflection. Secondly, when ethical reflection results in the wish to alter morality, it is good to know the implications: biological drives are, even in humans, psychologically hard to neglect! Let us therefore start with looking for a biological explanation of the taboo.

There are in chimpanzee or bonobo groups all kinds of social mechanisms which prevent group members from hurting, let alone killing, each other. For example young animals are not attacked; in a fight, subdued behaviour stops the fighting, etc. These mechanisms enhance the functioning of the group and thus also the evolutionary fitness of both group and individual. It is plausible that a rule like 'not killing a group member' is such a mechanism. These kinds of evolutionary rules are often preserved in the genes and so passed on to next generations. I hypothesise that the above mentioned 'taboo of killing humans' is derived from the evolutionary rule of 'not killing group members'. So a better description of the taboo is: there is a taboo on killing human group members.

The next interesting question is: what is a group member? I think that we do not have to pursue this (very interesting, but extensive) question here: a patient in our care or a member of our family certainly belongs to our 'group'. Killing them would be called 'murder' reflecting the wrongness of the act. So let's assume that the taboo on killing the patient can be biologically explained by the 'instinctive' rule of not killing a group member.

7.3 Ethical reflection on the taboo

We now have biological explanation for the taboo on killing our patient. But a biological explanation is not the same as an ethical justification. It is one thing to explain on biological grounds why we do not actively kill the suffering dementia patient, but it is something completely different to say it is *morally good* not to do it! To claim that 'it is in our genes and therefore it is good', does not count as sound ethical reasoning. It is a naturalistic fallacy: it derives a moral norm directly from an empirical fact. A moral norm needs a moral justification and not only a factual one.

References

Brody, B.A., 1973. Abortion and the sanctity of human life. American Philosophical Quarterly 10: 133-140.

De Tavernier, J., 2014. Morality and nature: evolutionary challenges to Christian ethics. Zygon 49: 179-181.

De Waal F., 2008. De aap en de filosoof. Uitgeverij Contact, Amsterdam, the Netherlands.

Encyclopaedia Britannica, undated. Taboo. Available at: http://www.britannica.com/topic/taboo-sociology.

Gunderson, A., 2004. Kantian view of suicide and end-of-life treatment. Journal of Social Philosophy 35: 277-287.

Kaldewaij, F., 2013. The animal in morality, justifying duties to the animal in Kantian moral philosophy. Questiones Infinitae LXXI: 219-221.

Koninklijke Maatschappij voor Diergeneeskunde (KNMvD), 2010. Code voor de dierenarts, KNMvD, Houten, the Netherlands.

Midgley, M., 1993. The origin of ethics. In: Singer, P. (ed.) A compendium to ethics. Blackwell Publishing Ltd., Malden, Oxford, UK, pp. 3-13.

Singer, P., 1980. Practical ethics. Cambridge University Press, Cambridge, UK.

Singer, P., 2011. Practical ethics. Cambridge University Press, New York, NY, USA.

Van Holsteyn, J. and Trappenburg, M., 1998. Citizens' opinions on new forms of euthanasia. A report from the Netherlands. Patient Education and Counseling 35: 63-73.

Van Wijlick, E. and Van Dijk, G., 2015. Dokters hikken soms tegen euthanasie aan. Medisch Contact 1: 16-19.

Part II.

Societal debates in the context of killing animals for animal disease prevention and control

8. Morality, morbidity and mortality: an ethical analysis of culling nonhuman animals

B. Mepham

Centre for Applied Bioethics, School of Biosciences, University of Nottingham, Sutton Bonington Campus, Sutton Bonington LE12 5RD, United Kingdom; ben.mepham@nottingham.ac.uk

Abstract

The fact that both humans and nonhuman animals utilise the world's natural capital means that conflicts of interest are ultimately inevitable. From an ethical perspective, omnivorous humans are obliged to manage those nonhumans they exploit for food in ways that they consider respect their rights and welfare; but all human moral agents (including vegans) also have responsibilities to ensure the ethical soundness of their actions that affect other humans and nonhumans alike. The case is often made that, in certain circumstances, taking everything into consideration, selective killing (culling) of nonhumans is an ethical requirement. This chapter seeks to examine the validity of that claim in several different contexts, by citing examples that refer to farm, wild and companion animals, in circumstances where there are alleged threats to human health and economic considerations, animal welfare and/or environmental sustainability. It is suggested that ethical deliberation on these issues in an era characterised by a constant flux in social, economic and cultural norms may be facilitated by employment of the *ethical matrix*. Use of this conceptual framework is exemplified here in considering the practice of culling badgers to abate the increasing incidence of bovine tuberculosis in dairy cattle.

Keywords: ecology, ethical matrix, veganism, zoonoses

8.1 Introduction

This chapter has three aims. It begins with consideration of some commonly-advanced reasons that allegedly provide the ethical justification in certain circumstances for culling nonhuman animals (hereafter, referred to simply as 'animals'), and following that, an overview is presented of a range of categories in which this justification is claimed to apply. The chapter concludes with an analysis of a specific instance currently due to be practised in England, the culling of badgers which is being performed with the object of reducing the incidence of bovine tuberculosis (bTB) in dairy cows. This latter part of the chapter suggests that public deliberation on the ethics of culling may be facilitated by employing the author's ethical matrix in this novel context.

*Franck L.B. Meijboom and Elsbeth N. Stassen (ed.) **The end of animal life: a start for ethical debate***
DOI 10.3920/978-90-8686-808-7_8, © Wageningen Academic Publishers 2016

8.2 Culling defined and justified

Despite the similarity of the two words 'culling' and 'killing,' they are etymologically distinct, with culling being derived from a word meaning 'collecting or selecting'. But in practice, with reference to animals, this amounts to selective killing, when this is clearly different from indiscriminate slaughter, or when it serves the main reason for the animal's existence, e.g. when the animal is destined to be killed to provide meat. In my subsequent use of the word 'culling,' the quality of 'selectivity' is always implied, although in common usage the distinction between culling and slaughter is sometimes ill-defined.

What in general are the reasons advanced for culling animals? It is a fact of life that the activities of all living organisms tend, over time, to have effects that are deleterious to living processes as a whole. Thus, they utilise components of the Earth's resources in ways that either pollute or deplete vital resources at rates exceeding natural replacement. When there is insufficient capacity to restore the initial conditions, this results in a substantial erosion of the Earth's natural capital – i.e. the stock of natural ecosystems that provides a sustainable flow of valuable goods or services – on which all life depends. Consequently, we are now experiencing increasingly serious conflicts of interests between humans, and between humans and animals, because the survival of all of us depends on having reliable access to adequate resources, especially as food.

It is a salutary fact that in the nonhuman world the sustainability of ecosystems is often highly dependent on predator-prey relationships. For example, the Serengeti in Africa is home to approximately 70 larger mammalian and 500 bird species, which in highly synergistic fashion exploit as food both the habitats composed of forests, swamps, grasslands and woodlands and several of the other animal species that inhabit the ecosystem. However, in the 'domesticated' regions of the world animal agriculture contributes significantly to the erosion of natural capital, because farm animals are both energetically inefficient in converting plants into animal products such as meat, milk and eggs, and because their unwanted products (e.g. greenhouse gases and excreta) often pollute the environment. Consequently, vegetarians argue that excluding meat (and in the case of vegans, all animal products) from the diet would both greatly reduce such problems and, crucially, respect animals' intrinsic value as sentient beings.

But the vegan position is challenged by those who argue (e.g. George, 2004) that exclusively plant-based diets are nutritionally inadequate for many people (e.g. young children, menstruating and lactating women, and many who are elderly and/ or unwell), whereas intensive arable management procedures (e.g. entailing use of pesticides, fertilisers and heavy farm machinery) inevitably kill many wild animals, either directly or indirectly (Davis, 2003). Moreover, in many developing countries the multiple roles of animals (in transport, traction, fuel and fibre provision, as well as supplying human food) make them key elements in families' domestic economy. Consequently, 'humans cannot live without exerting influences on other species. From an ecological standpoint, to live is to consume, to consume is to compete, to compete successfully is to out-compete, i.e. to work to the detriment of something else, be it plant or animal' (Scanlon, 1983). Arguably then, as participants in the global ecosystem, all humans act (directly and/or indirectly) as 'predators,' even though the origins of their food (e.g. when purchased from supermarkets) are often skilfully disguised by food processing and packaging.

Even so, animal agriculture frequently exacerbates certain problems in consequence of disease conditions that reduce the efficiency of food production from animals, reduce animal's welfare (sometimes fatally), and/or threaten (and sometime claim) human lives.

In short, it is arguable that the Earth's limited capacity to support human and animal life in acceptable ways often means that we sometimes encounter circumstances in which the culling of animals is ethically justified. In addition to such ecological factors, economic, political and legal considerations are also often invoked as motives for terminating animals' lives – however undesirable (or perhaps unacceptable) that solution might otherwise be. On the other hand, the worldview advanced by many vegans denies a necessity of killing animals for any morally justifiable reasons. There is thus a sense in which claimed ethical justifications for culling are direct challenges to vegan ethics; and the chapter may be regarded as an attempt to explore this fundamental ideological rift.

8.3 Justifications advanced for culling animals

It was indicated above that culling is generally invoked as a remedy for problems that arise when there are conflicts of interest between living organisms, human and animal. At the outset, it is perhaps worth considering the sorts of justification normally advanced in support of intentionally killing other *people* (Mepham, 2008). These include: (1) self-defence (e.g. if attacked by an assailant with apparently murderous intent); (2) 'just' wars (e.g. when inevitable losses of human life are considered morally justified in preventing worse crimes against humanity); (3) compassion (e.g. euthanasia of terminally ill people who prefer a gentle death to enduring extreme suffering); and (4) in certain countries, as perceived justified punishment for a capital crime. In all cases, these are grave decisions, which are ideally only arrived at, if at all, after deep reflection and circumspection. But at the other end of the zoological scale (or perhaps of sentience), many people believe it morally right to kill vermin (such as rats), few have scruples over killing insects if they are a 'nuisance,' and virtually everybody considers we are morally obliged to kill bacteria that cause infectious diseases.

Clearly, the animals to which the term culling is applied (mostly mammals, but also other vertebrates) lie somewhere in between these extremes; and it would be surprising if some of the motives that correspond to (1) – (4) above did not, sometimes, act as morally justifiable grounds for culling animals. On the other hand, according to a university extension service circular, if you keep sheep as a hobby 'culling may be important to you for aesthetic reasons' (Filley, 2009).

However, as in the case of humans, there would seem to be an ethical need to satisfy some strict criteria. Thus, arguably, culling should only be performed: (1) for a sound, perhaps even *vital*, reason; (2) after having seriously considered all possible options for addressing the problem; and (3) humanely, as far as possible avoiding pain and distress (in both animals and people). Of course, even general agreement on such conditions does not guarantee that well-meaning moral agents will necessarily agree on each case in which culling is practised or proposed. Unpacking the above three criteria is an objective of deliberations based on the ethical matrix, to be discussed below.

It is not the intention here to survey the range of methods employed, but characterising the main approaches, albeit very briefly, is important because the way culling is to be achieved has a strong influence on whether it should be practised at all.

Methods vary in many respects, e.g. their efficacy, impacts on animal welfare, difficulty and expense in performing, need for training of operators, requirement for specialised equipment and public acceptability. Some animals can be killed relatively painlessly by injection of lethal chemicals or by inhalation of poisonous gases; some are shot by marksmen (the acceptability of which obviously depends in part on human skill), but where animals are trapped before being shot death may be instantaneous although possibly preceded by stress and injury.

Humane slaughter of large mammals may entail stunning with a captive-bolt, followed by sticking (exsanguination) or pithing (destruction of the central nervous system); but for some species (e.g. sheep and pigs) stunning entails electric shock treatment, which in the case of poultry is achieved by immersing the bird's head in electrified water. Grandin (2006) has pioneered techniques of animal slaughter designed to minimise stress. Male day-old chicks, unwanted in the egg industry, are usually killed by instantaneous maceration, neck dislocation or gassing. However, the feasibility of any technique depends largely on the animal's size, situation (e.g. feral or on a farm) and the availability of expertise and equipment.

8.4 Circumstances suggesting the need to cull animals

The aim here is to provide an overview of a (by no means comprehensive) range of circumstances in which the practice of culling is claimed to be ethically justified.

	

It is useful to list these under four categories. For each category, one example is described in greater detail than others in order to characterise the basis of the alleged justification. Some conditions fall into more than one category: and some issues (e.g. the actual culling process) are discussed more fully in some sections than in others.

8.4.1 Physical and psychological threats

Such threats can be posed:
- to people, e.g. when dangerous animals escape from zoos or farms, dogs attack children, or rats raid food stores;
- to animals when suffering from untreatable injury, e.g. when birds become road accident victims or horses are injured in racing events, or when subject to poor welfare, e.g. because born with serious deformities or neglected by incompetent owners.

Culling unwanted pets

A prominent example of this category consists of animals, such as cats and dogs, that are taken to, or rescued by, animal shelter organizations. Often originally bought by people unaware of the personal commitments that pet ownership entails, culling is the predominant fate of such animals because their total far exceeds the numbers of people offering to adopt them. The result is that, e.g. in the USA, millions of cats and dogs, the majority of which are healthy animals, are humanely killed every year, mostly by injection of barbiturates.

It is often argued that humane killing is the best solution: (1) for the individual animal, thereby relieving it of unnecessary suffering; (2) for (former) owners who are incapable of providing adequate care; and (3) human communities, which would otherwise be prey to the destructive activities of excessive numbers of feral animals. Palmer (2008) argues that this solution ignores a 'relational approach', that takes account of the degree to which some animals can live fulfilled lives largely independently of human support. For her, 'the question then arises of whether it is better to live a life of ferality, providing it is not one of interminable agony, than to be painlessly killed'. On the other hand, if we have, collectively, produced pets that are highly dependent on human support, e.g. by the way they have been selectively bred, it is arguable humans have ethical responsibilities to ensure their continued fulfilled lives.

But in other cases, where the primary purpose of culling is to relieve suffering, of the animal concerned or others who might be seriously affected by its behaviour, it is often appropriate to consider the act of culling to be a form of euthanasia (RCVS, 2012). In most cases of injury to wild animals (e.g. in road accidents) a lethal injection is the preferred method.[1] Firearms are used only rarely, and when this is impracticable (e.g. because the animal is aggressive) a procedure involving a 'captive bolt', as used in slaughterhouses, may be employed.

8.4.2 Zoonotic diseases

These are diseases in which infections are transmissible between species, and, most significantly, from animals to humans. Examples are *bovine spongiform encephalopathy* (BSE) and rabies, both of which can be fatal for humans, and also have serious consequences for conspecifics and related species. What is perhaps less widely appreciated is the pervasive and critical influence of zoonotic diseases that originate in wildlife species, which may well be the source of 'almost all emerging diseases of humans' (Maccallum and Hocking, 2005). Examples are HIV, Ebola virus, severe acute respiratory syndrome and avian influenza.

Culling is a standard response to zoonotic disease outbreaks, but in the case of wild life species, because the animals are free-ranging, the problems of this practice are clearly much greater than with farm animals. Vaccination has been proposed as an alternative, but this is also logistically complex, difficult to target effectively, and of uncertain efficacy. Another option is release of infectious agents that will kill animals acting as a reservoir of the diseases, a notorious instance of which was *myxomatosis* to control the rabbit population in Australia. Genetically modified pathogens are currently being developed to achieve this objective (Maccallum and Hocking, 2005).

Bovine spongiform encephalopathy

The rationale for culling is exemplified here by BSE. Often known as 'mad cow disease', this is a fatal cattle disease that produces spongy degeneration in the animal's brain and spinal cord. First identified in the UK in 1996, the causative agent is a prion protein, which remains viable even at the high temperatures to which the body tissues of other animals were exposed before being fed to cattle as meat and bone meal (MBM). This 'rendering' process, based on the economic motive of recycling tissues of potentially high nutritive value, was probably the cause of material from sheep infected with scrapie (a closely related, but apparently non-contagious prion disease) entering the human food chain.

BSE is most readily transmitted to some humans by consuming meat from carcases contaminated with tissue from the cow's brain, spinal cord or digestive tract. In humans, the disease is called 'new variant Creutzfeld-Jakob Disease' (nvCJD), which, unlike CJD itself, particularly afflicts younger people. To date, over 200 people have

---

[1] Euthanasia statement of the Royal Society for the Prevention of Cruelty to Animals (RSPCA), 2012. Available at: http://tinyurl.com/owcnrns.

died from nvCJD, mostly in the UK; but this is much lower than the millions some earlier predicted would die. Over 180,000 cattle have been infected with BSE, and 4.4 million culled in the eradication programme (National Archives, 2006).

Culling of cattle was justified in terms of the urgent need for drastic action at a time when scientific understanding was very poor – perhaps a prime example of adopting the precautionary principle at a time of great uncertainty. Although understanding of the aetiology of BSE remains incomplete, it is a widespread opinion of experts in this field that the disease has now virtually died out.

8.4.3 Economic threats

These are usually associated with microbial diseases of farm animals, which threaten farmers' incomes because they result in animal deaths or low productivity and/or fertility, e.g. BSE, foot and mouth disease (FMD), and bTB. But economic factors are also considered as acceptable grounds for killing male chicks in the egg industry, and in arable systems for killing pests eating food crops (e.g. rabbits and pigeons). The key role of agriculture in national and international agricultural, food, environmental and cultural contexts also means that the economic impacts of large-scale disease outbreaks (epizootics) can be highly significant, even when no threat is posed directly to public health. The latter was the case with a recent FMD epizootic in the UK, the following consideration of which exemplifies this category.

Foot and mouth disease

After the BSE outbreak, the UK was again plunged into a crisis by an FMD epizootic, which mainly affected sheep, but also cattle and pigs. Described as 'the most devastating disease of farm animals in the world', the outbreak which began in 2001, involved the slaughter of millions of animals, by methods that, because of the panic situation, were often inhumane. Although technically a cull, the deaths of millions of animals (including all that were within 3 km zone of an outbreak) were elements in the, so-called mass slaughter programme (MSP).

During the course of the epizootic, I gave an account of the UK outbreak, and the measures being taken to contain it (Mepham, 2001). I challenged the basis of the MSP on the grounds that although it was ostensibly based on utilitarian reasoning, it was almost impossible to conceive how its main objective, regaining disease-free status for the UK as quickly as possible, could outweigh the harm inflicted on animals, the damage to commerce (in terms of the virtual closure of the rural tourist industry), to the environment (in terms of pollution from burning and buried carcases) and to human sensibilities from witnessing, often ill-managed, slaughter procedures.

The vaccination option discussed in the paper, and more extensively by others (e.g. Woods, 2004) is now a central feature of the UK government's FMD control strategy, which seeks *inter alia* to 'minimise the number of animals culled either for disease control purposes or to safeguard animal welfare,' and to 'minimise adverse impacts on animal welfare, the rural and wider community, the public, and the environment' (DEFRA, 2011). Progress is being made on improved vaccines, and a cheaper, potentially lower-risk vaccine for cattle is in prospect (Grubman *et al.*, 2010).

In retrospect, the crisis emphasises the need for stricter restrictions on off-farm-movements of animals, preparedness for worst-case scenarios, and fuller consideration of the political, social and economic ramifications of such disease outbreaks (Jensen, 2004; Mepham, 2004; Murphy-Lawless, 2004).

8.4.4 Environmental damage

Animals living in essentially feral conditions can threaten group survival of conspecifics, and other animal and plant species (e.g. deer in Scotland). The case of elephants in an African wildlife reserve is discussed here in order to illustrate the issues involved.

Elephants in the Kruger National Park, South Africa

Over a century ago, elephants in Africa were virtually wiped out because they were hunted and killed for their ivory and skin. But the establishment of reserves by relocation of animals ultimately led to excessive growth of populations, such that the Kruger National Park (KNP) is now home to about 15,000 elephants, and countless other species of fauna and flora. Unfortunately, the voracious appetites of elephant herds are highly detrimental to the stability of the ecosystem, such that the future survival of many living species, including the elephants themselves, appears to be seriously threatened.

This threat led to calls for a sustained cull of elephants, which many considered the obvious solution to the problem. Indeed, in order to keep the elephant population of the KNP to below a ceiling of 7,000, a total of over 16,000 elephants were culled between 1967 and 1994. However, in 1994 pressure from animal rights groups led to a moratorium on future killing while the policy of culling underwent public debate and a full scientific assessment. The expert panel's conclusions suggest that the problem is extremely complex, but, crucially, that 'there is no compelling evidence for the need for immediate, large-scale reduction of elephant numbers in the KNP' (Owen-Smith *et al.*, 2006). Even so, other knowledgeable experts challenge this conclusion.

The complex interactions between animal and plant species, some of which are synergistic while others are predator-prey relations, together with climatic variables and the influence of factors such as fires, floods and droughts – all produce ecological changes in states of temporal and spatial flux. These cause marked fluctuations in biodiversity, so that disturbances may not return to an original state for decades, if indeed they ever do so. In the face of such uncertainty, even if one were to adopt an approach ostensibly guided by the precautionary principle, the correct strategy would by no means be clear. For aiming to reduce elephant numbers to prevent loss of biodiversity might be one interpretation of the principle, whereas avoiding reducing elephant numbers to maintain biodiversity might be another: yet they are clearly consistent with culling and non-culling, respectively.

The expert panel concluded *inter alia* that: (1) the previously ceiling of 7,000 should not be construed as a carrying capacity; (2) there is no benchmark against which to judge an ideal state of KNP vegetation; and (3) culling alone may actually make matters worse, not least because of elephants' behavioural responses (Owen-Smith *et al.*, 2006). On the other hand, possible strategies might entail: (1) increasing the calving interval, through reduced fertility, e.g. by restricting surface water supplies; (2) selective, regional culling; and (3) immuno-contraception in small, enclosed reserves (Druce *et al.*, 2011).

From the ethical perspective argued by Regan (1983), that lays emphasis on respect for animal sentience, killing contravenes each animal's rights, but for those who prioritise ecological integrity (e.g. Callicott, 1980), individual animals' interests must be considered in a holistic context, which may, regrettably, entail 'therapeutic culling'. The central problem is that humans' desire to escape the rigours of 'the wild' could only be successful if an impermeable barrier were to be established between nature and human culture. But in a shrinking world this becomes increasingly difficult, even if desirable.

However, the case for 'therapeutic culling' poses further challenges, because stripped of its clinical associations, this amounts to 'hunting', an activity some people consider intrinsically unethical. Varner (2011) has defined 'therapeutic hunting' as that form 'designed to secure the aggregate welfare of the target species across generations, the health and/or integrity of its ecosystem, or both'. But while the consequence might be deemed worthy, in practice, the motives of the hunters might be based on the purely selfish desires of winning trophies and prestige among fellow 'sportsmen' and/or experiencing the thrill of killing other living (and in this case very large) sentient beings. Such motives clearly offend Kant's stricture, that rightful actions should be performed out of duty, and that any personal rewards gained nullify the morality of the actions.

A useful, alternative, categorization of reasons cited for culling wild animals in the UK has recently been provided by Fowler-Reeves (2007) viz. the animals are regarded as aliens, pests and/or predators – designations that the author considers are often highly questionable. Indeed, the grounds advanced for killing animals commonly appear to be based more on emotion and prejudice than on reason.

8.5 Postmodernism and prima facie principles

The diversity of the issues raised above might appear to elude any overarching principle that would suggest grounds for a uniform ethical approach to culling. The basis on which it is commonly justified varies from urgent measures to protect human and animal health; to economic concerns affecting individuals, industries or nation states; and to long-term ecological considerations that could crucially affect global environmental sustainability. The ethical challenges also relate to the relative weight to be assigned to humanity, individually or collectively; to nonhumans as sentient beings or as species; and to the biosphere as a whole, now and in future. Moreover, such values are inextricably embedded in the 'irreducible uncertainty of ecological systems' (Maccallum and Hocking, 2005) and the doubts that surround attempts to act in accordance with the precautionary principle. And yet, as often in political decision-making, precautionary actions are preferable to precautionary paralysis.

A traditional approach to normative ethics aspires to provide a sound philosophical basis for decisions on how we should behave, by appealing to theories that will guide us in seeking to achieve the 'right' or the 'good' solutions. We might only rarely have the strength of character, depth of altruistic intent or moral courage to pursue the actions that these theories propose, but it is considered that we would know, a least 'on paper ' what we *should* do – a disparity philosophers in ancient Greece called *akresia*. But the central problem we now face is that, in our postmodern era, the assumption that ethics simply entails applying to real-life circumstances the schemes derived from rigorous, principled, 'top-down' reasoning, no longer seems capable of guiding us to sound decisions.

There are several possible reasons for this, for example: (1) the challenge presented by new scientific understanding to the authority long associated for many people with religious insights; (2) the clash of different cultures in our increasingly globalised world; (3) many people's increased economic and political power; and (4) increasing awareness of the impending ecological threats to global sustainability due to anthropogenic activities. The combination of such factors tends to induce a sense of resigned inertia. For as, sociologist Bauman (2008) has observed, in our 'liquid modern, individualised consumer society,' the speed with which new phenomena 'burst into public awareness and disappear from view (prevents) the experience from crystallizing, settling and solidifying in attitudes and behavioural patterns, value syndromes and worldviews'.

Thus, in seeking to make ethically justifiable decisions about whether, and if so, how humans should kill nonhuman animals it seems crucial that we are cognizant of the relentless flux of events, the wide scope of interests that demand attention and the relevance of the social and political context in which such decisions need to implemented.

8.5.1 The ethical matrix

My claim is that the framework of the ethical matrix (EM) may well provide an effective means of addressing these issues by facilitating open, but appropriately structured, ethical deliberation. The matrix approach to addressing ethical issues in a social context builds on the notion of *prima facie* principles advanced by Ross (1930), and applied in medical contexts by Beauchamp and Childress (e.g. 1994). Several publications illustrate the use of the ethical matrix (hereafter, EM), but two accessible introductory accounts are freely available on websites (Mepham. 2008; Mepham and Tomkins, 2003).

The EM is based on principles that are grounded in the common morality, i.e. the norms of ethical behaviour and belief widely accepted in a society, which incorporate *prima facie* principles based on utilitarian, deontological and Rawlsian theory (Mepham, 2008). Thus, the EM's advantage over 'top down' theory is that it starts with 'common sense' perceptions of ethics and then seeks to refine these in the light of both deliberative reasoning and a focus on a specific issue. It is interesting that the survey of public attitudes to culling reported by Cohen *et al.* (2009) identifies what the authors term 'fundamental moral attitudes' that bear a strong identity with the specifications assigned to the *prima facie* principles featuring in the EM. Thus, even in the 'liquid' society we all inhabit, some broad ethical standards persist, and provide us with a sound basis for future action.

The principal claims for the value of the EM when used in public consultations and by committees, were summarised by Kaiser and Forsberg (2001) as follows. It: (1) is liberal regarding the approach to be adopted, enabling it to be read equally as a utilitarian or deontological approach; (2) provides substance for ethical deliberation; (3) translates abstract principles into concrete issues that concern those unfamiliar with ethical theory; (4) facilitates extension of ethical concerns, e.g. to include democratic decision-making; and (5) captures the basic fact that although options considered are likely to affect different stakeholders in different ways so that they are likely to reach different decisions as to the best course of action, the aim is to find an optimal solution in the light of these conflicting interests.

8.6 The case of badger culling in England

The EM's potential value in ethical analysis is illustrated here by considering the disease bTB, which afflicts dairy cattle. Currently, the disease costs the UK over £100 million p.a., mostly in compensation to farmers for slaughtered cattle. Although bTB is principally contracted by transmission from infected cattle, badgers can also become infected with the bacterium *Mycobactrium bovis*, and convey it from farm to farm. This led to the Government's decision to implement a badger cull programme in England, initially scheduled to begin in 2012 in two areas of the south west, but, subject to satisfactory progress in these two pilot schemes, ultimately spreading to over 40 sites. This example conforms to category 3 (economic threats), as identified above. Under the Bern Convention, badgers are a legally protected species, but culling is permitted in certain cases to prevent disease transmission.

Opponents of culling (e.g. the Badger Trust[2]) advance arguments based on animal rights and culling's poor efficacy. Thus, the perturbation caused is certain to displace badgers from one location to another, thus transferring, or possibly exacerbating, the problem. The Independent Scientific Group on Cattle TB (ISG) reported in 2007 that careful evaluation of their and others' data, indicated that badger culling could make no meaningful contribution to cattle TB control in Britain. Moreover, the report suggested that 'weaknesses in the cattle testing regimes ... contribute significantly to the spread and persistence of the disease' and that vaccination, of both cattle and badgers, might be effective in controlling and reducing the severity of bTB.

Such deficiencies were confirmed by a European Commission audit (EC, 2012) which identified 'a fragmented system of controls, involving a number of responsible bodies' which made it difficult to ensure that basic practices, such as cleaning and disinfection of vehicles and markets, were 'carried out in an effective way'. It was however, acknowledged that effective vaccination programmes need to be developed with urgency. Ironically, the Government's scientific advisers do not question the ISG's estimates of the cull's efficacy in reducing bTB, but consider that the small effect (perhaps, at best a 15% reduction in bTB) will be significant (DEFRA, 2012b).

Economic considerations inevitably play a role, because trapping badgers before shooting (ensuring a clean kill) is more expensive than 'free shooting', the practice to be employed in the English cull. The Department of Food and Rural Affairs (DEFRA) (2012a) issued 'best practice guidance for the shooting of free-ranging badgers,' which is to be limited to trained people who have been granted a licence certifying their competence. But in October 2012, when the cull was due to begin, the Government reported that the most recent estimates of badger numbers in the pilot areas were

---

[2] Available at: www.badger.org.uk/Home.aspx.

much greater than previously envisaged, leading the National Farmers Union to call for a delay until summer 2013 because of the increased costs of implementing an effective cull.

Further frustrating the plans, in October 2012, in a non-binding vote that had been triggered by over 150,000 emails from UK citizens calling for it, MPs rejected the culling policy by 147 votes to 28, calling instead for vaccination, improved testing and biosecurity measures.

The pros and cons of culling versus vaccination have been debated extensively for many ears, and no attempt is made here even to summarise the debate. However, it is noteworthy that the UK Government's recent decision to adopt culling in England reverses an earlier opinion, which was based on much authoritative scientific evidence (including the ISG report); while in 2012 the Welsh Assembly abandoned an earlier Assembly decision to cull badgers in favour of a vaccination-based policy.

It is in this context that use of the EM may facilitate sound decision-making, in that it explicitly structures deliberation according to *prima facie* ethical principles, rather than resorting to an adversarial contest based on ill-defined premises. With reference to Table 8.1, the following impacts seem likely to be important in reaching a sound decision on the ethics of badger culling. In each case my comments are based on the assumption that alternative approaches (entailing effective vaccination, regulation and testing) *might* collectively constitute a more ethically justifiable alternative.

8.6.1 Use of the ethical matrix to facilitate ethical analysis of badger culling

Points at issue are identified by noting the specific cell (e.g. cattle wellbeing or badgers autonomy) of the EM to which they pertain (Table 8.1). Not only does this ensure that deliberation is appropriately focused, but also that the analysis is structured in terms of respect for the underlying ethical principle. The following synopsis of the ingredients of the ethical analysis is in no sense comprehensive, and within the confines of this chapter is intended merely to suggest an agenda for ethical deliberation. The claims made are not comprehensively referenced.

8.6.2 Dairy cattle

In the short term, cattle in the cull area may be less likely to contract bTB, but in the longer term, due to 'perturbation' the problem may simply be transferred to surrounding areas (cattle wellbeing and autonomy; Table 8.1). According to a DEFRA report (2007), 'cattle vaccination has potential benefits to reduce prevalence, incidence and spread of bTB (cattle wellbeing), but because it is not 100% effective it 'cannot be

Table 8.1. Ethical matrix: suggested specifications of prima facie principles with reference to badger culling and bovine tuberculosis.[1]

Respect for	Wellbeing	Autonomy	Fairness
Cattle (with, or at risk of contracting, bovine tuberculosis)	1. Healthy life, within safe environment 2. Humane treatment on farms and at slaughter	1. Ability to express natural behaviour 2. Respect for animal rights	Absence of unreasonable restrictions (compared with cattle in non-cull areas)
Badgers (targets of the culling process)	1. Healthy life, within safe environment 2. Humane treatment in life and/ or at culling	1. Ability to express natural behaviour 2. Respect for animal rights	Absence of unreasonable restrictions (compared with badgers in non-cull areas)
Dairy farmers (at risk of personal distress and/or economic disadvantage)	Human safety and confidence in efficacy of procedures employed to address the problem	Freedom to act in accordance with conscience in animal management	Absence of adverse discrimination, physical and mental threats and burdensome regulation
Managers (government regulators, marksmen, vets)	Human safety and confidence in efficacy of procedures employed to address the problem	Freedom to act in accordance with conscience in animal management	Absence of adverse discrimination physical and mental threats and burdensome regulation
Public (as citizens and consumers of animal products)	Insignificant undesirable effects e.g. physical or zoonotic disease threats , or culturally offensive practices	Life's normal activities and choices unaffected by procedures adopted to address problem	Absence of adverse discrimination in cull areas by comparison with unaffected areas or countries

[1] For each cell of the ethical matrix, the relevant principle is specified in terms of *respect* for the principle. The aim is then to assess whether, and how much, each principle is respected or infringed by culling *versus* alternative strategies (see text), and how, collectively, these impacts influence a final decision.

used to define disease-free status'. However, in relation to improvements consequent on a vaccination programme, infringements of fairness due to culling would appear likely to be substantial (cattle fairness; Table 8.1).

In an open letter to the government, more than 30 eminent animal disease experts claimed that 'licensed culling risks increasing cattle TB rather than reducing it' (cattle wellbeing) and, accusing ministers of 'failing to tell the truth,' demanded immediate abandonment of the killings.[3]

---

[3] Available at: http://tinyurl.com/klljop6.

8.6.3 Badgers

The decision to recruit marksmen (many of whom may be the farmers themselves) to kill badgers by free-shooting raises the risk of some badgers escaping but with a serious injury, leading to painful deaths (badgers wellbeing; Table 8.1). Moreover, the regulations require that when an injured badger is believed to have taken refuge in a sett, the sett must not be dug into or interfered with in any way. If sett interference was considered in the badger's best interests, obtaining a licence for this would probably entail a significant delay and further suffering. Shooting is clearly inconsistent with animal rights (badgers autonomy; Table 8.1), but tests also revealed that 80% of the over 30,000 badgers killed between 1975 and 2004, were free of bTB (badgers fairness; Table 8.1) (Fowler-Reeves, 2007).

8.6.4 Dairy farmers

In the short term, cattle in the cull area may be slightly less likely to contract bTB, raising farmers' wellbeing and autonomy (Table 8.1); although compared to improvements consequent on a vaccination programme impacts on fairness (farmers fairness; Table 8.1) might not be substantial. But in view of widespread disquiet over the cull, adverse public reactions seem likely, possibly involving violent protest or verbal abuse directed at farmers (farmers' wellbeing).

8.6.5 Managers

In the interests of brevity, this interest group includes politicians authorising the cull, marksmen and officials implementing it, and veterinary surgeons monitoring it. But, in practice, there might be merit in focusing on the distinctive issues confronted by the separate elements of this group. Reasonable doubts about personal competence may be distressing for marksmen and others involved in culling (managers' wellbeing; Table 8.1), while failure of personnel to comply on conscientious grounds might adversely affect their job security (managers' autonomy; Table 8.1). In view of widespread public disquiet over the cull, adverse public reactions seem likely, and may well (as openly warned) involve violent protest and attempts to sabotage culling (managers' fairness; Table 8.1) as extremist groups join the protests.

8.6.6 General public

The cull programme is unlikely to impact directly on public health because there is virtually no risk of people contracting bTB from pasteurised milk and dairy products (public wellbeing; Table 8.1). But public disquiet, accentuated by militant and vociferous opponents, has stirred considerable criticism of culling; while those engaging in obstructive forms of protest may be put at risk of serious physical harm (public wellbeing). However, as culling will largely be confined to discrete rural areas, a significant impact on most people's autonomous activities (public autonomy) or sense of fairness (public fairness; Table 8.1) may be less evident.

8.6.7 Summary

The above brief analysis is intended merely to suggest, rather than define, an approach to this issue, whereas authentic public deliberation would involve contributions from a range of experts and perspectives, and ample opportunity for questioning and reflection. Although the EM approach aims to achieve disaggregation of the issues on the basis of ethical principles concerned, as they impact on each interest group affected, it also exposes claims allegedly based on scientific data to criticism of a more sociological nature. For example, according to Grant (2009) 'mythical constructions of the badger have shaped the policy debate' so that 'relevant evidence was incomplete and contested' and 'alternative framings of the policy problem were polarised'. Consequently, the normal techniques of stakeholder management through co-option and mediation have been marginalised.

Use of the EM could change the public focus from markedly polarised disputation between proponents and opponents of culling (in which conceding even a single point to one's opponent is often deemed undesirable) to one entailing consideration of points on which a measure of consensus might readily be achieved. Arguably, the approach described will continue to be valid, even though by the time this book is published there may have been further significant developments on the issue.

8.7 Challenges

Among other considerations, reasoned use of the EM will demand:
- *Of those espousing an absolutist animals rights philosophy*: that they are able to justify their concern for the rights of individual animals, when this will inevitably lead to aggregate harm and infringement of the rights of other living beings.
- *Of those who support culling*: that they are able justify this when appropriate use of alternatives (such as combination of vaccination, animal translocation, amended EU legislation and tighter safety regimes) might achieve more effective results with less infringement of human and animal rights and welfare.

Effective use of the EM may facilitate participatory deliberation on these challenging issues and the attainment of ethically sound decisions, perhaps involving prudent compromises.

The specific case of bTB illustrates how the EM might structure fruitful deliberation in accordance with overriding *prima facie* ethical principles. Indeed, the EM appears to have attracted considerable attention (cf. internet Google searches) from people addressing a wide, and heterogeneous, range of concerns. This suggests that there is much interest in a generic approach to decision-making that is grounded in ethical principles. It is perhaps fitting that, now in the later stages of my academic career, I plan to conduct an analysis of EM use to discover how, and how effectively, it has been used.

References

Bauman, Z., 2008. The art of life. Polity Press, Cambridge, UK, 152 pp.

Beauchamp, T.L. and Childress, J.F., 1994. Principles of biomedical ethics. Oxford University Press, Oxford, UK.

Callicott, J.B., 1980. Animal liberation: a triangular affair. Environmental Ethics 2: 311-338.

Cohen, N.E., Brom, F.W.A. and Stassen, E.N., 2009. Fundamental moral attitudes to animals and their role in judgment: an empirical model to describe fundamental moral attitudes to animals and their role in judgment on the culling of healthy animals during an animal disease epidemic. Journal of Agricultural and Environmental Ethics 22: 341-359.

Davis, S.L., 2003. The least harm principle may require that humans consume a diet containing large herbivores and not a vegan diet. Journal of Agricultural and Environmental Ethics 16: 387-394.

Department for Environment Food and Rural Affairs (DEFRA), 2007. Options for vaccinating cattle against bovine tuberculosis. Available at: http://tinyurl.com/99wx6bz.

Department for Environment Food and Rural Affairs (DEFRA), 2011. Foot and mouth disease control strategy for Great Britain. Available at: http://tinyurl.com/kp8rvup.

Department for Environment Food and Rural Affairs (DEFRA), 2012a. Controlled shooting of badgers in the field to prevent the spread of bovine TB in cattle. Available at: http://tinyurl.com/kuc5p55.

Department for Environment Food and Rural Affairs (DEFRA), 2012b. Bovine TB: the Government's approach to controlling the disease. Available at: http://tinyurl.com/3mfxoo5.

Druce, H.C., Mackey, R.L. and Slotow, R., 2011. How immunocontraception can contribute to elephant management in small, enclosed reserves: Munyawana population as a case study. PLoS ONE: e27952.

European Commission (EC), 2012. Final report of an audit carried out in the United Kingdom from 05 to 16 September 2011) in order to evaluate the operation of the bovine tuberculosis eradication programme. European Commission, Health and Consumers Directorate-General. Available at: http://tinyurl.com/nkvddz3.

Filley, S., 2009. Culling ewes. L&F Circular no. 0602. Oregon State University, Roseburg, OR, USA.

Fowler-Reeves, K., 2007. With extreme prejudice: the culling of British wildlife. Animal Aid, Tonbridge, UK, 60 pp.

George, K.P., 2004. A paradox of ethnic vegetarianism: unfairness to women and children. In: Sapontzis, S.F. (ed.) Food for thought: the debate over eating meat. Prometheus Books, Amherst, NY, USA, pp. 261-271

Grandin, T., 2006. Progress and challenges in animal handling and slaughter in the U.S. Applied Animal Behaviour Science 100: 129-139.

Grant, I.W., 2009. Intractable policy failure: the case of bovine TB and badgers. The British Journal of Politics and International Relations. 11: 557-573.

Grubman, M.J., Moraes, M.P., Schutta, C., Barrera, J., Ettyreddy, D., Butman, B.T., Brough, D.E. and Brake, D.A., 2010. Adenovirus serotype 5-vectored foot-and-mouth-disease subunit vaccines: the first decade. Future Virology 5: 51-64.

Independent Scientific Group (ISG), 2007. Bovine TB: the scientific evidence. Final report of the Independent Scientific Group on Cattle TB. Available at: http://tinyurl.com/3jy456j.

Jensen, K.K., 2004. BSE in the UK: why the risk communication strategy failed. Journal of Agricultural and Environmental Ethics 17: 405-423.

Kaiser, M. and Forsberg, E.M., 2001. Assessing fisheries – using an ethical matrix in a participatory process. Journal of Agricultural and Environmental Ethics 14: 191-200.

Mccallum, H. and Hocking, B.A., 2005. Reflecting on ethical and legal issues in wildlife disease. Bioethics 19: 336-347.

Mepham, B., 2001. Foot and mouth disease and British agriculture: ethics in a crisis. Journal of Agricultural and Environmental Ethics 14: 339-347.

Mepham B., 2004. Farm animal diseases in context. Journal of Agricultural and Environmental Ethics 17: 331-340.

Mepham, B., 2008. Bioethics: an introduction for the biosciences. Oxford University Press, Oxford, UK, 440 pp.

Mepham, B. and Tomkins, S., 2003. Ethics and animal farming: an interactive web program. Availabele at: www.ethicalmatrix.net.

Murphy-Lawless, J., 2004. The impact of BSE and FMD on ethics and democratic process. Journal of Agricultural and Environmental Ethics 17: 385-403.

National Archives, 2006. BSE Inquiry: the report. Available at: http://tinyurl.com/nqqdte3.

Owen-Smith, N., Kerley, G.I.H., Page, B., Slotow, R. and Van Aarde, R.J., 2006. A scientific perspective on the management of elephants in the Kruger National Park and elsewhere. South African Journal of Science 102: 389-394.

Palmer, C., 2008. Killing animals in animal shelters. In: Armstrong, S.J. and Botzler, R.G. (eds.) The animal ethics reader, 2nd edition. Routledge, London, UK, pp. 570-582.

Regan, T,. 1983 The case for animal rights. University of California Press, Berkeley, CA, USA.

Ross, W.D.,1930. The right and the good. Oxford University Press, Oxford, UK.

Royal College of Veterinary Surgeons (RCVS), 2012. Euthanasia of animals. Available at: http://tinyurl.com/pgtgp7g.

Scanlon, P.F., 1983. Humans as hunting animals. In: Miller, H.B. and Williams, W.H. (eds.) Ethics and animals. Humana Press, Clifton, NJ, USA, pp. 204-205.

Varner, G., 2011. Environmental ethics, hunting and the place of animals. In: Beauchamp, T.L. and Frey, R.G. (eds.) The Oxford handbook of animal ethics. Oxford University Press, New York, NY, USA, pp. 854-876.

Welsh Assembly, 2012. Environment minister announces programme of badger vaccination. Available at: http://tinyurl.com/7r747m8.

Woods, A., 2004. Why slaughter? The cultural dimensions of Britain's foot and mouth disease control policy, 1892-2001. Journal of Agricultural and Environmental Ethics 17: 341-362.

9. Public moral convictions about animals in the Netherlands: culling healthy animals as a moral problem

N.E. Cohen[*] and E. Stassen
Wageningen University, Department of Animal Sciences, Adaptation and Physiology Group, P.O. Box 338, 6700 AH Wageningen, the Netherlands; nncohen8@gmail.com

Abstract

In this chapter the dynamics of public moral convictions about animals in the Netherlands are described in the context of animal disease epidemics. A change has taken place in these convictions, due to a shift in the relational value of animals and the emergence of new animal practices in Dutch rural countryside. This played a major part in the public resistance against the large scale culling of healthy animals in recent animal disease epidemics. The chapter describes and analyses the moral values at stake and argues that differences in the choice and weight of these values were at the heart of this conflict. New policy acknowledging the relevance of these values is briefly discussed.

Keywords: culling, value of life, risk of harm

9.1 Introduction

Animals have always played an important part in human society. In the history of our relation with animals, animals provided us with food and clothes, guarded the house, and worked on the land. Over the years a shift has taken place, in which animals are increasingly valued as well for their own sake as a companion and as living creatures with an intrinsic value (Franklin, 2006; Rollin, 2007).

De Cock Buning *et al.* (2012) in a Dutch study among 2011 respondents showed that people have a strong emotional relationship with animals (70%) and that animal welfare is valued highly. The study also shows that people consider a number of animal practices, such as livestock farming, as intrinsically related to animal welfare concerns.

Our human-animal relationship is also ambiguous (Cohen *et al.*, 2009; Pagani *et al.*, 2007; Rutgers and Swabe, 2003). Animals with which we have strong emotional ties and are 'visible' (such as companion animals) are often valued over animals that have an instrumental value to us and are 'invisible' (such as laboratory or production animals). De Cock Buning describes an alienation from commercial animal practices on the one hand and a further personal relation on the other. This implies that our

*Franck L.B. Meijboom and Elsbeth N. Stassen (ed.) **The end of animal life: a start for ethical debate***
DOI 10.3920/978-90-8686-808-7_9, © Wageningen Academic Publishers 2016

relationship is not the same for all animals all the time, and should be viewed in its context. In this friction, the values of animal welfare and life are variable.

Not only our changed relationship with animals but also the emergence of new animal practices in the Dutch countryside have exerted their influence on how we relate to animals. The rural area is no longer dominated by livestock farmers, but now includes other animal practices as well, such as backyard animal keeping, animals in nature reserves, recreation with animals (such as horseback riding), and care farms. In these practices, the human-animal relation is personal as well as instrumental. Backyard animal keepers keep their animals for breeding special or rare species, company, grazing and other non-commercial purposes. They have interests other than those of commercial farmers. Also the rural area had become attractive to city people, who had moved to the country to experience animals as part of nature.

These developments have their influence on the public moral convictions about the right or wrong way to treat animals from the perspective that animals are living beings that can be harmed by our actions. These conviction are shaped by a multitude of social, cultural, and religious influences (Heleski *et al.*, 2006; Pagani *et al.*, 2007), personal experience (Miura *et al.*, 2002), and knowledge about the mental capacities of animals (Bekoff, 2007; Knight and Barnett, 2008).

Convictions develop and become practical when used in a real life situation. They are brought to life, shaped, reshaped, re-valued or solidified in a public debate on a moral issue in a specific circumstance and context. Then, the conviction once again becomes embedded in the moral history of an individual or that of a society. This means that a conviction can exists in a theoretical form and in a practical form, and is best described by the dynamic interaction between the two forms. The implication is that we should attempt to learn more about people's theoretical convictions, because they do exist in some form and need to be understood. However, for a comprehensive understanding of the dynamics of convictions and their diversity in society, we also need to learn about their role in the context of a practical animal issue.

A number of recent epidemic outbreaks of animal diseases in the Netherlands provided the animal issue required and therefore were used as a case study to study the dynamics of public convictions in the Netherlands (Cohen, 2010).

9.2 Culling animals during the outbreak of infectious diseases: a case study

9.2.1 Non-vaccination policy

In the early nineties the European Union adopted a non-vaccination policy to control highly contagious animal disease epidemics, which included a stamping-out of the disease by culling infected and healthy animals within a radius of 1-3 kilometres from the source of the infection (KNAW, 2002; Mepham, 2004; Woods, 2004). In 1997-98, 2001, and 2003, Europe faced epidemic outbreaks of classical swine fever, foot and mouth disease, and highly pathogenic avian influenza, respectively. Especially the Netherlands were hard hit.

In these epidemics millions of infected and healthy animals were culled. The latter were included because these animals could still be carriers of the disease in question. This was from an economic perspective preferable to vaccination. Furthermore, the culling strategy did not distinguish between commercial and non-commercial animal practices, which meant that not only production animals were culled, but backyard animals, and animals in in nature reserves as well.

9.2.2 Public resistance

The culling caused major public resistance throughout the whole of Dutch society and made clear that some major changes in our relationship with animals had taken place, which had remained unnoticed up till then (Cohen *et al.*, 2009; Huirne *et al.*, 2002; NVBD, 2004; RLG/RDA 2003, 2004; Van den Berg, 2002; Van Haaften and Kersten, 2002). This resistance was based on a number of issues.

9.2.3 Animal welfare

The animal welfare problems encountered (not only in the Netherlands but in the United Kingdom as well) were a major topic in the public discussion (NVBD, 2004; Van den Berg, 2002). The scale of the slaughter to be performed within a limited time-frame, combined with a control strategy which was not adequate to deal with the scale of the epidemic, led to animal welfare problems involving cases of improper handling, killing, stunning, and transport of animals. Handling, restraint and killing methods in the field are very different from those in slaughterhouses. Concern was expressed over the unsuitable conditions for on-farm slaughter and inappropriate killing methods. These problems were exacerbated by the fact that handling and slaughter were sometimes in the hands of unskilled personnel not accustomed to working in disease control/field situations. Movement restrictions due to a transport ban were reported to cause major animal welfare problems. Overcrowding caused physical problems in rapidly growing poultry and aggression and cannibalism in pigs.

9.2.4 Social and ethical implications

A number of studies were performed to clarify the impact of the stamping-out strategy on animal keepers and the general public. In a study, conducted by Huirne *et al.* (2002), a questionnaire was sent to 662 respondents among the Dutch general public. The results showed that the foot and mouth epidemic had left a deep impression, especially with respect to the way the animals had been culled and disposed of (73%). Other domains of concern were the emotional and financial impact on the farmers, the way the crisis had been handled by the authorities, the isolation of the farmers, and the fact that animals were no longer seen in the countryside. The preferred strategy during a future outbreak was vaccination of all animals (70%) and isolation (54%), while a majority (72%) dismissed culling healthy animals to stop the spread.

Rutgers and Swabe (2003) performed a study into the societal and moral acceptability of the killing of kept animals. A total of 1,939 respondents selected from the Dutch general public participated in the study, and in-depth interviews were performed with 43 experts. The majority (84%) of the respondents were of the opinion that the culling of healthy animals is morally unacceptable when the control strategy is based exclusively on economic motives, which are governed by European trade policies serving the livestock industry, which can be fairly described as large-scale and focused on the export market. The prevailing view was that the control strategy values economic interests over the lives of living creatures. It was not, however, considered unacceptable to kill animals for food production.

These findings were corroborated by Stafleu *et al.* (2004) and De Greef *et al.* (2006), who described the relationship between farmers and their animals. These farmers felt that the farmer and the animal each have a role to fulfil in the world as providers of high quality food. In this view, the killing of a healthy animal for the production of food is considered acceptable because it is the natural life cycle of a production animal; but the culling and destruction of healthy animals as a control measure during an epidemic for economic reasons is considered unacceptable because the 'natural function' of the animal would not yet have been fulfilled.

In the study of De Cock Buning *et al.* (2012) 45% of the respondents considered culling problematic from an animal welfare point of view, and were against the culling of healthy animals.

9.2.5 Moral values at stake

The public debate concentrated on a number of moral values: the intrinsic and relational value of an animal's life, the duty to treat animals well (to care for their health and well-being and to protect them against harm), and the autonomy of the animal keepers to care for their animals and their business as they saw fit, within the restrictions of food safety and public health regulations These values are moral values because they concern the right or wrong treatment of animals, based on the fact that animals are living beings and can be harmed by our actions. It became clear that these values were not the priority values for the authorities.

First, to the authorities, the value of an animal's life was interpreted as its economic value (to a farmer, the livestock sector, or the country). The interests at stake were basically economic, therefore the loss of a number of animals compared to the benefits for the sector and the country as a whole, was justified in an economic sense (Mepham, 2001) and was thought to be sufficiently covered by a financial compensation per animal. This compensation was based on their economic value, but other values were not included, such as the value of special and rare breeds, kept by backyard animal keepers or zoos especially for this reason.

In Dutch society (Cohen *et al.*, 2010) the value of an animal's life often referred to the intrinsic value of the animal in its own right as a living being and the value of the personal and emotional relationship between people and their animals. The morally laden terms 'right to life' or 'respect for life' were used to express this opposition.

Second, an important issue was the 'duty to treat animals well' (Crispin *et al.*, 2002), which for the animal keepers was the core responsibility to their animals. This was in their view a moral duty: people deliberately choose to keep and confine animals, and therefore are responsible for their health and well-being, and have a duty to protect them against harm. In their view they were forced to act against this moral duty, because economic duties to the nation prevailed.

Third, 'autonomy' to the individual keeper or animal practice, meant to be at liberty to act according to one's own convictions to properly care for and protect their animals and their enterprises. To the authorities autonomy was a value that could be outweighed by national interests (Meijboom *et al.*, 2009). This justified the decision to take control of the private domain of the animal keepers, rendering them powerless to stop the slaughter men entering their premises and harming and culling their animals. At a different level, animal keepers had been denied the choice to vaccinate their animals to protect them against these diseases.

respondents were younger and female, had more contact with animals in private life or work, and lived in larger towns or cities.

The position that humans are superior to animals (A) was mostly based on the opinion that animals do no possess rationality, and can't distinguish between right and wrong. Those that considered humans and animals to be equal (B) emphasised the similarities between the two: the fact that both are living beings, are sentient, and are both equally important in the ecosystem.

Both A and B held the position that animals have value in their own right as well as having an instrumental value for humans. The majority of A and B held the position that people have a duty to care for and protect all animals.

The three most important values in support of this position was the opinion that animals are living beings, are sentient, and are important in the ecosystem. Therefore, the fact that this value is relevant to all animals means that a duty to care is not dependent on a certain degree of sentience.

The majority of A and B held the position that all animals have a right to life, which was mostly based on the fact that they are living beings and are important in the ecosystem. Animals were recognised as having an interest to live and fulfil the goal of their life independently of the interests of people.

9.3.2 Part two

In the second part of the study more insight was obtained about the role of convictions in judgement on the culling of healthy animals during the outbreaks of animal disease epidemics (the practical form). The A and B respondents were asked to give their judgement in four cases: culling to stop the disease from spreading, to safeguard the export position of a country, to protect human health (eye infection) and to protect human life.

Most A respondents agreed with the culling to stop a disease from spreading and most A and B respondents agreed with the culling to protect human life. Of both profiles 50% partly (dis)agreed with the culling to safeguard the export position. To protect human health 41% of the A profile and 45% of the B profile partly (dis)agreed with the culling. In all four cases, more A respondents than B respondents agreed with the culling and more B respondents disagreed or partly (dis)agreed.

In judgement (Table 9.2) 'animal life is valuable' was the core argument against culling. Differences in judgement were based on differences in the valuation of animal life in

Table 9.2. Mean rating of the argument 'animal life is valuable; therefore these healthy animals should not be culled' by the A and B profile.[1]

Judgement	Rating by the respondents who disagreed with the culling		Rating by the respondents who partly (dis)agreed with the culling		Rating by the respondents who agreed with the culling	
Profile	A	B	A	B	A	B
Animal life is valuable, therefore healthy animals should not be culled to:						
stop the disease from spreading	7.1[*]	7.9[*]	5.4[*]	6.0[*]	2.6[*]	3.5[*]
safeguard the export position of a country	6.1[*]	7.0[*]	4.7[*]	5.7[*]	3.0	3.5
protect human health (eye infections)	6.1[*]	7.4[*]	5.0[*]	6.2[*]	3.4[*]	4.2[*]
protect human life	5.6[*]	8.0[*]	5.1[*]	6.1[*]	3.0[*]	4.2[*]

[1] The rating on a scale between 1-10 reflects the importance given to the judgement; [*] P<0.05.

a specific context. Animal life was valued lower when human life is at stake. Animal life was valued higher when purely economic values are at stake.

9.4. Risk of harm and policy

The public resistance against the culling was not incidental, but the result of the then existing moral convictions in Dutch society. This justifies a new perspective on an animal disease policy that has such a strong impact on animals and people. Policy should be able to deal with the demands of current public morality about the intrinsic value of animal life, which is the value of life in its own right. This means that life itself is valuable, irrespective of its value to people, such as for food, company, or recreation. This value has gained a prominent place in the public morality, in order to gain more support in Dutch society. As a starting point, it should be acknowledged that in epidemics harm is done to all: the animals, the animal keepers, and society as a whole.

Though the nature of animal practices may differ, their interest not to be harmed by an epidemic is the same. Morally justifiable policy aims to equally distribute risk-of-harm between the stakeholders. This requires a reassessment of the boundaries of stakeholders' autonomy and their responsibilities towards others. From a moral perspective, giving priority to economic risk only is morally wrong, because it runs contrary to a consideration of all values.

Given the diversity of moral convictions, any prevention and control policy includes certain risk-of-harm to those involved (Meijboom *et al.*, 2009). For instance, a non-vaccination policy reduces the economic risk to the livestock sector, but is a risk to backyard animal keepers who are denied the choice to protect their animals from a serious disease by vaccination.

Current policy has already acknowledged these convictions and risks-of-harm by shifting towards a policy based on preventive measures (www.rijksoverheid.nl), such as quarantine measures, and by means of biotechnical solutions, such as biochips and resistance building by immunisation, improved early warning systems and risk assessment in other countries. At the same time better vaccines are being developed. Revised contingency plans for classical swine fever and foot and mouth disease allow vaccination under certain circumstances.

Policy needs to be developed at the European level, because animal disease policy is an international issue. In is unlikely that all member states share the same moral values. The model described is a valuable instrument to learn more about the diversity of convictions between the member states and about the morally acceptable approach to prevent and control these diseases.

References

Bekoff, M., 2007. The emotional lives of animals. New World Library, Novato, CA, USA.

Cohen, N.E., 2010. Considering animals: moral convictions concerning animals and judgement on the culling of healthy animals in animal disease epidemics. PhD thesis, Wageningen University, Wageningen, the Netherlands.

Cohen, N.E., Brom, F.W.A. and Stassen, E.N., 2009. Fundamental moral attitudes to animals and their role in judgment: an empirical model to describe fundamental moral attitudes to animals and their role in judgment on the culling of healthy animals during an animal disease epidemic. Journal of Agricultural and Environmental Ethics 22: 341-359.

Crispin, S.M., Roger, P.A., O'Hare, H. and Binns, S.H., 2002. The 2001 foot and mouth disease epidemic in the United Kingdom: animal welfare perspectives. Revue Scientific et Technique International Office of Epizoonotics 21: 877-883.

De Cock Buning, T., Pompe, V., Hopster, H. and De Braauw, C., 2012. Denken over dieren. Dier en ding, zegen en zorg. VU University Amsterdam, Athena Institute, Amsterdam, the Netherlands. Available at: http://tinyurl.com/oua87w7.

De Greef, K., De Lauwere, C., Stafleu, F., Meijboom, F., De Rooij, S., Brom, F. and Van der Ploeg, J.D., 2006. Towards value based autonomy in livestock farming? In: Kaiser, M. and Lien, M.A. (eds.) Ethics and the politics of food. Wageningen Academic Publishers, Wageningen, the Netherlands, pp. 61-65.

Franklin, J.H., 2006. Animal rights and moral philosophy. Columbia University Press, New York, NY, USA.

Heleski, C.R., Mertig, A.G. and Zanella, A.J., 2006. Stakeholder attitudes towards farm animal welfare. Anthrozoös 19: 290-307.

Huirne, R.B.M., Mourits, M., Tomassen, F., De Vlieger, J.J. and Vogelzang, T.A., 2002. Verleden, heden en toekomst. Over de preventie en bestrijding van MKZ, report LEI. LEI, the Hague, the Netherlands.

Knight, S. and Barnett, L., 2008. Justifying attitudes towards animal use: a qualitative study of people's views and beliefs, Anthrozoös 21: 31-42.

Koninklijke Nederlandse Akademie van Wetenschappen (KNAW), 2002. Bestrijding van mond- en klauwzeer, 'stamping out' of gebruik maken van wetenschappelijk onderzoek? KNAW, Amsterdam, the Netherlands. Available at: http://tinyurl.com/qh4sxvc.

Meijboom, F.L.B., Cohen, N.E., Stassen, E. and Brom, F.W.A., 2009. Beyond the prevention of harm: animal disease policy as a moral question. Journal of Agricultural and Environmental Ethics 22: 341-359.

Mepham, B., 2001. Foot and mouth disease and British agriculture: ethics in a crisis. Journal of Agricultural and Environmental Ethics 14: 339-347.

Mepham, B., 2004. Farm animal diseases in context. Journal of Agricultural and Environmental Ethics 17: 331-340.

Miura, A., Bradshaw, J. and Tanida, H., 2002. Childhood experiences and attitudes towards animal issues: a comparison of young adults in Japan and the UK. Animal Welfare 11: 437-448.

Nederlandse Vereniging tot Bescherming van Dieren (NVBD), 2004. Reactie van de Dierenbescherming op de evaluatie van de vogelpestcrisis 2003 n.a.v. het rapport van het Bureau Berenschot en de reactie van het kabinet. NVBD, the Hague, the Netherlands.

Pagani, C., Robustelli, F. and Ascione, F.R., 2007. Italian youths' attitudes toward and concern for animals. Anthrozoös 20: 275-293.

Raad voor het Landelijk Gebied en Raad voor Dieraangelegenheden (RLG/RDA), 2004. Dierziektebeleid met draagvlak, advies over de bestrijding van zeer besmettelijke dierziekten. Deel 2 – onderbouwing van het advies, report RLG 03/8 RDA 2004/01.

Raad voor het Landelijk Gebied en Raad voor Dieraangelegenheden (RLG/RDA), 2003. Dierziektebeleid met draagvlak, advies over de bestrijding van zeer besmettelijke dierziekten. Deel 1 – advies, report RLG 03/8 RDA 2003/08.

Rollin, B.E., 2007. Cultural variation, animal welfare and telos. Animal Welfare 16: 129-133.

Rutgers, L.J.E. and Swabe, J.M., 2003. Het doden van gehouden dieren. Utrecht University, Utrecht, the Netherlands.

Stafleu, F.R., Lauwere, C.C. De Greef, K.H., Sollie, P. and Dudink, S., 2004. Boerenethiek. eigen waarden als basis voor een 'nieuwe' ethiek. Een inventarisatie, verkennende studie. NWO, the Hague, the Netherlands.

Van der Berg, B., 2002. Evaluatie van het welzijn van dieren tijdens de mond- en klauwzeercrisis in Nederland in 2001. Nederlandse Vereniging tot Bescherming van Dieren, the Hague, the Netherlands.

Van Haaften, E.H. and Kersten, P.H., 2002. Veerkracht. Report of Alterra, no. 539. Alterra, Wageningen, the Netherlands, 213 pp.

Woods, A., 2004. Foot and mouth disease in Britain: the history. Centre for the History of Science, Technology & Medicine. A manufactured plague: the history of foot-and-mouth disease in Britain. Routledge, Abingdon, UK, 240 pp.

10. Premature culling of production animals; ethical questions related to killing animals in food production

M.R.N. Bruijnis[1], F.L.B Meijboom[2] and E.N. Stassen[1]
[1]Wageningen University, Department of Animal Sciences, Adaptation and Physiology Group, P.O. Box 338, 6700 AH Wageningen, the Netherlands; [2]Utrecht University, Ethics Institute, Janskerkhof 13a, 3512 BL Utrecht; the Netherlands; m_bruijnis@hotmail.com

Abstract

The aim of this chapter is to analyse the importance of longevity in relation to the welfare of production animals. I hypothesize that the concept of longevity helps to support the moral intuition that premature culling of animals is a moral wrong. The analysis shows that the interpretation of the concept of animal welfare is important for decisions on whether or not to cull animals, but also for the measures that should be taken to prevent premature culling. This is illustrated by two examples in animal production, one example relating to dairy cattle and the other to breeding sows. These two types of farming have in common that in these practices animals are necessary to produce products, yet this production does not require– the animal itself to be killed. My proposal is to accept the view on animal welfare according to which longevity is accepted as an independent moral argument. Acceptance of this view substantiates the intuition that premature culling of animals is a moral wrong, because it shows that we have additional reasons to give the interests of animals more weight. In order to respect this view, some common practices in animal farming will become the subject of debate, as illustrated in the two cases.

Keywords: animal welfare, longevity, production animals

10.1 Introduction

The provision of safe and affordable food has been a leading driver of development in animal farming. In the 20th century, evolving technological innovations made it possible to develop efficient farming systems. Farms keep large numbers of production animals, which considerably reduces the input of labour and the use of space. Efficiency is further improved by the use of medicines (e.g. vaccines and antibiotics), balanced diets and genetic improvement, generating high production levels per animal. Many animal production chains have been intensified, from poultry to dairy cattle to pig farming (CBS, 2014; FAO, 2012). Some of these production systems, such as poultry farming, deal with such large numbers of animals that decisions on managing the animals and their health and welfare involve flock decisions and do not include the

Franck L.B. Meijboom and Elsbeth N. Stassen (ed.) The end of animal life: a start for ethical debate
DOI 10.3920/978-90-8686-808-7_10, © Wageningen Academic Publishers 2016

individual animal. Other types of animal production still include decisions at the individual level, for example dairy farming and sow breeder farms. Here, each animal represents enough value in itself to be considered individually.

The acceptability of animal production systems is a subject of debate, as there are diverse views on the necessity of producing animal products and how it should be done. People's opinions differ as to how they value the position of animals, the importance of welfare and what welfare means. All farming systems create health and welfare problems to a certain degree, as there are space restrictions, limitations on an animal's ability to express natural behaviour, etc. Moreover, possible improvements in animal health and welfare need to be weighed against other values relating to animal production systems, such as good working conditions, farm profitability, environmental impact and risks to public health.

In addition to the welfare of production animals during their lifetime, the end of animals' lives is also a subject for discussion. For animals which produce the end product directly, such as broiler chickens, fattening " and beef cattle reared for meat, it is usually clear when they will be slaughtered, i.e. when they have reached a certain weight. If an animal has severe health and welfare problems before reaching the slaughter weight, it will be culled prematurely. For animals that are not the end product themselves, the end of life is less clear. These include breeding sows that produce fattening pigs, laying hens that produce eggs and dairy cows that produce milk. Important factors in the decision whether or not to cull an animal are how well the animal is still performing and what its health status is. For laying hens, the time for slaughtering the animals is determined in advance for the whole flock. The hens are slaughtered at around 80 weeks of age, because they have become economically non-viable as the number and quality of eggs produced will have decreased by then. The whole flock is replaced at once, with the exception of animals that have died or have been culled prematurely due to health or welfare problems. For production systems where culling decisions still involve individual animals, the decision is more difficult. In practice, the decision whether or not to replace an animal depends on productivity, health status, animal welfare, availability of replacement animals, etc.

Welfare, including health, is often considered to be relevant to the animal, given its capacity to experience pain and pleasure (EC, 2009). There is less agreement about the animal's interest in the duration of its life (i.e. longevity), and to what extent these notions contribute to animal welfare. Nonetheless, we can observe that, based on clear moral intuitions, people experience the premature culling of production animals as a moral wrong (Balcombe, 2009; Cohen *et al.*, 2009; De Cock Buning *et al.*, 2012; Rutgers *et al.*, 2003).

The aim of this chapter is to analyse the importance of longevity in relation to the welfare of production animals. I hypothesize that the concept of longevity can help to support the moral intuition that premature culling of animals is a moral wrong. The hypothesis will be applied to examples relating to dairy cattle and breeding sows. These two types of farming have in common that they involve animals where decisions still include the individual animal and the animals produce products which do not require the animal itself to be killed.

In the following paragraphs I will first discuss some points that are fundamental to this chapter. These include an explanation of the assumption that animals such as dairy cows and breeding sows have a life worth living, and what is meant by premature culling. Next, I will analyse the moral intuition that premature culling is a moral wrong by analysing the concept of animal welfare as a biological and normative concept in relation to premature culling and longevity. This will be applied to cases in dairy farming and sow breeding. I will end with some conclusions.

10.2 A life worth living

For all animal farming systems, the basic question can be asked: do the animals in such production systems have a life worth living? Whether animals kept for production have a life worth living is discussed in more detail in other chapters of this book. Some views, such as 'respect for life' (Taylor, 1986) and 'respect for individuality' (Regan, 1983), consider keeping animals for production to be highly problematic, if not totally unacceptable. The main question of this chapter – whether longevity of production animals substantiates the intuition that premature culling is a moral wrong (and whether longevity is a constitutive element of animal welfare) – cannot be answered with these views, because they already see the keeping of animals for food production as problematic. An evaluation of the acceptability of animal farming is beyond the scope of this chapter. Although the quality of life of different types of production animals can be debated, in this chapter I start from the assumption that the animals have a life worth living. I make this assumption as it is acknowledged in the EU that animals are sentient beings (EC, 2009) and guidelines have been formulated to protect livestock (The Council of the EU, 1998). In the Netherlands, too, the intrinsic value of animals is acknowledged, which means that animals are recognized as moral subjects that have value as an end in itself. Acknowledging animals as sentient beings and granting them intrinsic value implies respect for the interests of the animals, including their welfare. Minimum requirements (based on the Five Freedoms[1]) are set to ensure that animals receive care (based on the duty of care). These include the requirement that animals receive sufficient and good quality feed and water, that they be protected from pain and disease, that prolonged stress be avoided and that

---

[1] The Five Freedoms were formulated by the Brambell committee in 1965 and formulated more broadly by the Farm Animal Welfare Council (http://tinyurl.com/o3l5v5w).

animals be able to express natural behaviour. It can be questioned whether some of these requirements can be met in practice. Sows in farrowing pens, for example, will not be able to express natural behaviour. However, such discussions are also beyond the scope of this chapter.

10.3 Premature culling

To be able to discuss examples of premature culling in animal production, it is important to know what I mean by culling and when culling is premature. Culling means the removal of an animal from the farm, in general for slaughter. Premature culling implies the decision to end the life of an animal prematurely, i.e. before the end of a normally intended productive life. This might be due to disease or health problems which cause welfare problems or which cost the farmer money and time for treatment, resulting in lower production results. In addition to health problems, replacement with young stock that has a higher production potential can be another reason for premature culling.

In this chapter I will not discuss moral problems related to the slaughtering of animals in general, because there is general public acceptance of slaughtering animals for food (De Cock Buning *et al.*, 2012; Rutgers *et al.*, 2003). The focus of this chapter is the role of longevity before animals reach their intended life span as production animals. The age that an animal could reach in theory is not considered suitable as a reference for the lifespan of production animals.

10.4 Premature culling: does it relate to animal welfare?

As stated in Bruijnis *et al.* (2013b), both biological and normative assumptions define to what extent an animal has interests, what these interests are and what they imply for our dealings with animals. Schmidt (2011) endorses this with the following well-formulated statements: 'animal welfare science and animal ethics highly depend on each other'. And 'the background of values underneath every welfare theory is essential to pursue animal welfare science. Animal ethics can make important contributions to the clarification of underlying normative assumptions with regard to the value of the animal, with regard to ideas about what is valuable for the animal, and with regard to actions that should follow from the results of animal welfare science'.

Thus, the combination of one's biological and normative value assumptions leads to specific views on animal welfare and – as I claim – also on how one judges the relationship between premature culling and animal welfare. By combining three views on animal welfare based on biological knowledge (i.e. biology, ethology, and physiology) with four views based on moral norms relating to animals and their welfare, it has been shown that the impact and importance of premature culling depends on one's view of animal welfare (Bruijnis *et al.*, 2013b). The biologically based views on animal welfare can be divided into three overlapping and complementary views: animal welfare as a matter of health and functioning, feelings, and natural living (Fraser *et al.*, 1997). These views have different biological starting points, according to which subscribing to one view does not rule out supporting a different one. For example, if the view concerning feelings is supported, it does not rule out holding the view regarding health. These biological views can be related to four normative views about animal welfare that indicate the moral importance of animal welfare and our duties towards animals. The normative views (based on, e.g. De Greef *et al.*, 2006; Nussbaum, 2006; Regan, 1983; Rollin, 2004) assume that sentient beings should: (1) be able to function well; (2) feel well; (3) be able to satisfy preferences; and (4) be able to develop according to species-specific needs and capabilities. Combining these biological and normative views leads to three different interpretations of the importance of longevity for animal welfare. These three interpretations (Figure 10.1) will be discussed in the following paragraphs.

10.4.1 Longevity as an indicator for animal welfare

The role of longevity as an indicator for animal welfare relates to the normative view of functioning well. This interpretation relates to the ability of animals to be healthy and, consequently, to be able to function well. In this view, welfare implies that the animal is free from disease and is able to function well, for example to grow and to

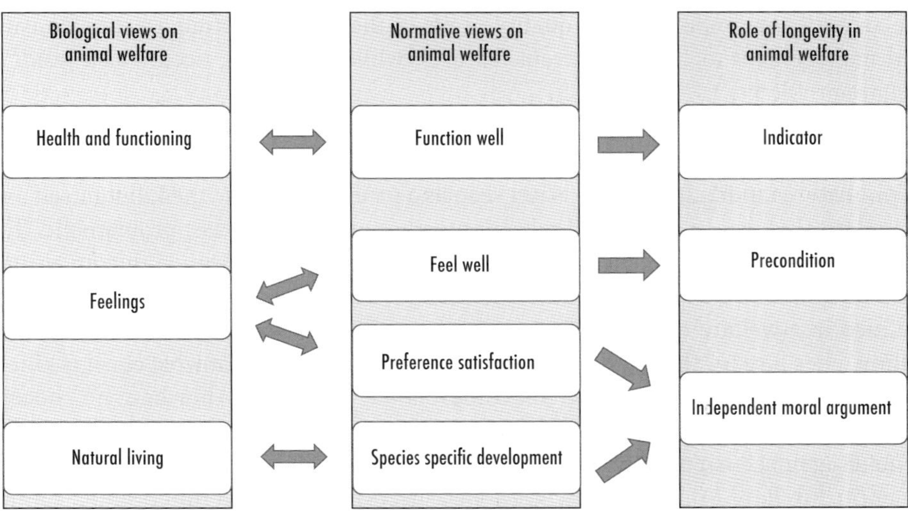

Figure 10.1. Relation of the biological and normative views on animal welfare to the role of longevity in animal welfare based on these views (Bruijnis et al., 2013b).

reproduce. Farmers in particular refer to this view on welfare when, for instance, they argue that if animals are healthy and producing well, it means that they are enjoying good welfare (De Greef *et al.*, 2006). This definition of animal welfare is rather limited. A short lifespan usually indicates serious health problems during the animal's life. A premature cull is therefore often seen as an indicator of reduced animal welfare (e.g. Broom, 2007). However, culling animals before the end of the intended lifespan is not necessarily an indicator of impaired welfare during life, as healthy animals are also culled prematurely, for instance when dairy cattle are replaced by young stock with higher milk production potential.

Evaluating animal welfare in livestock farming based on the interpretation of longevity as a welfare indicator might result in counterintuitive conclusions about animal welfare. To illustrate this, I will look at a situation where many animals in a herd are culled due to production diseases. On the one hand, the premature culling might indeed be the preferred option in such a situation if the animals are really suffering pain. The decision to kill the animals might thus be in the interests of the individual animal. This reasoning also includes aspects of feelings and illustrates that premature culling can be for the sake of an animal (see next paragraph). On the other hand, when premature culling is used to 'solve' health problems, a low average age of the herd will be the result. This low average age is an indicator for health and welfare problems. However, it might appear that such a herd has no health and welfare problems, because a herd-level welfare assessment will detect relatively few animals with a health problem at the moment of assessment. In reality, though, this so-called positive result is achieved without improving the health and welfare experienced by the animals (i.e. improving the quality of life of the animals through better housing and management to cure or prevent disease), while the killing occurs *as a result of* welfare problems.

Based on the health and functioning views only, 'solving' health problems by culling, or premature culling in general, is not seen as a problem because the killing itself and the relatively low age of the animals are mostly not interpreted as problematic. The argumentation usually includes the following steps. First, it is argued that an animal on a farm has to be killed anyway. The second step is the claim that if the animal is killed in a welfare-neutral way (i.e. without pain and stress), there is no harm in killing it because for the animal it does not matter whether and at what age it is killed. It is often argued that animals do not comprehend extra life, as they have no concept of time and lack any awareness of their future, and therefore cannot weigh future benefits against current misery (cf. Lund and Olsson, 2006; Rollin, 2006). Such a view on animal welfare, which only includes the aspects of health and functioning, cannot explain the moral intuition that premature culling is a moral wrong.

10.4.2 Longevity as a precondition for animal welfare

The role of longevity as a precondition for animal welfare refers to the view which concerns the ability to feel well. Treating animals' feelings as the key element of animal welfare is often accompanied by the moral claim that animals, as sentient beings which can feel pain and pleasure, should be granted moral status and be the subject of moral concern. This starting point has been defined in the Treaty of Lisbon, where it is stated that animals are sentient beings and that one needs to pay full attention to their welfare requirements (EC, 2009). This EU Treaty shows that animals and their interests ought to be taken seriously. In line with this, Yeates (2009) states: 'one condition for an issue being a welfare issue is that its assessment as an issue involves (or ought to involve) an evaluation of states with regard to an animal's interests'. In the practice of animal production, welfare issues arise because animal interests are at stake.

Speaking about the interests of animals is not a value-free act. Starting animal welfare discussions from the point of view of animal interests perfectly shows the relationship between the biologically and normatively based elements of animal welfare. Different views have been expressed, for instance those of Balcombe (2009) who starts from a standpoint of hedonistic utilitarianism and argues: 'Pleasure has moral import for practices like factory farming and laboratory research, for it amplifies the moral burden of depriving animals the opportunity to lead fulfilling enjoyable lives'. From a similar normative perspective, Bradley (2009) states: 'There is no good reason to discount the badness of death for an animal. If an animal would have had a good life, then killing it is bad for it, even if it cannot contemplate its future'. Furthermore, Sapontzis (1987: 166-170) has already argued that life is important for living things that have interests. Without continuing to live, the animal cannot fulfil its interests. From this perspective, a premature death not only implies depriving an animal of its life, but it also 'deprives it of a future in which it could pursue its interests' (Haynes, 2008: 54).

If the feelings of animals and the ability to feel well are a starting point, then being alive is a precondition for animal welfare. Depriving an animal of a positive future harms it, because it means depriving it of its good state of welfare (Balcombe, 2009). In this view, longevity can *contribute* to the animal's quality of life. Consequently, this view considers lifespan as relevant in the moral assessment of a practice of animal use, because the duration of life directly influences the ability to have positive experiences. Taking the feelings of animals as the starting point therefore does help somewhat in explaining the intuition that premature culling is a moral wrong, because being alive is a precondition for welfare and culling prevents the animal from enjoying a good state of welfare.

However, it is not the best idea to strive for an animal to have a longer life in all cases. Animals can be euthanized or culled prematurely because they are in pain or experience discomfort and stress (if there is no chance of improvement or if recovery will be a lengthy and painful process). A longer life does not necessarily improve animal welfare; it can also impair animal welfare, especially when ageing brings increasing health problems. Based on a prima facie responsibility to kill an animal that would otherwise have a life worth avoiding (Yeates, 2009), it is necessary to know when no treatment or improvement is possible, in order to protect the animal from unnecessary and severe suffering before it is killed.

10.4.3 Longevity as an independent moral argument for animal welfare

Preference satisfaction

The previous section showed that emphasizing the normative view focusing on the ability to feel well does not necessarily imply that longevity plays a role as an independent moral argument in welfare discussions. The emphasis is on feelings during life and a longer life might result in greater welfare, which makes longevity a medium for being able to feel, but does not directly relate to welfare. In order to see longevity as a constitutive element of animal welfare, we have to use another normative view relating to the feelings of animals, namely preference satisfaction. In short, this implies that the moral good entails maximizing preference satisfaction, and that animal welfare holds that satisfying animal preferences will result in positive feelings. An animal can have a whole range of preferences to satisfy, including the preference for survival. If this latter preference is at stake, then (a) longevity should be included in the ethical assessment of animal welfare (cf. Singer, 1993) and (2) longevity can then be seen as a welfare issue (independent moral argument), since killing the subject results in the frustration of its preference and negatively affects its overall welfare.

Those who interpret welfare mainly in terms of functioning and health offer counterarguments in response to this reasoning. They argue, for instance, that the importance of the desire to stay alive is only one possible preference, which needs to be balanced against an animal's other preferences and which would thus reduce the importance of longevity as an element of animal welfare. However, many essential elements of animal welfare need to be weighed against other interests. Furthermore, the preference for staying alive and living a long(er) life is often related to consciousness and to the capacity to comprehend concepts such as life, death, and the future (e.g. McMahan, 2002). Not everyone is convinced that animals have a level of consciousness such that they are able to have a preference for staying alive. On the one hand, this leads to empirical questions about whether it is possible to determine scientifically whether animals have a future orientation and, if so, whether they prefer to stay alive. The answers to these questions are still uncertain, but DeGrazia (1996), for example, argues that sentient animals have desires and that some of them have a sense of time in terms of memory (sense of the past) or anticipation (sense of the future). He emphasizes that the level of having desires and a sense of time can be different from that of humans but that the level is not decisive for acknowledging its moral importance. The question is whether or not there is some perception of time. If there is, the animal is not stuck in the present and will have an interest in more than just present experiences. Furthermore, based on empirical findings, there are indications that some animals have the capacity for a future orientation (e.g. Clayton *et al.*, 2003). On the other hand, even without conclusive empirical evidence, there is a moral reason to treat animals, such as cows and pigs, as having the capacity to have preferences regarding their future. I will use two arguments to substantiate this moral claim. First, having a future orientation implies a capacity to compare states at different moments. Comparison of different states by animals is frequently used in animal research and animal welfare discussions. Yeates (2009) expresses this appropriately in the statement: 'It follows from this comparative nature of (at least much of) welfare assessment that the value of a welfare state may depend not only on what that state included in itself, but also on what states would otherwise be present'. This principle of the comparison of animal welfare states at different moments can be illustrated by the practice of tail docking in pigs. Docking the tail at a young age, which is a painful procedure, prevents future pain due to tail biting. Thus, a possible violation of animal welfare in the future is the reason for a painful procedure in the life of many young piglets. Thus, again in line with Yeates (2009), comparing different states of welfare includes more than the present state of the animal; it also concerns what has been and what will be. The assumption that animal welfare is more than just the feelings and interests of the animal in the present makes it possible to state that future welfare and future interests will be infringed when the animal is killed. This explanation at least confirms the importance of feelings and the role of feelings as a precondition, but it is also a way of understanding the future preferences of an animal in the process of comparing preferences. Second, because of the possibility that animals have some future orientation, I use precautionary reasoning to give animals the benefit of the doubt. Even if we take the empirical uncertainty with respect to animals' consciousness seriously, the aspect of preference satisfaction helps to substantiate the moral intuition that premature killing includes a moral wrong. In this context, longevity can serve as an independent moral argument in the animal welfare debate.

Species-specific development

In the public debate, animal welfare is not only framed in terms of health and feeling, but also includes the aspect of natural living. This means the ability to express natural behaviour and fulfil (behavioural) needs that are important to – and characteristic of – that species. On this point there has been a clear change in the perception and evaluation of animals. Cohen *et al.* (2007) state: 'The relationship between man and animal has in recent years evolved from a purely functional relationship in which the animal is valued mostly for its instrumental utility to humans towards one in which respect for the value of the animal as a being in its own right plays a significant role'. For instance, the cow's need to explore, the chicken's need to have the opportunity to take a dust bath, or the pig's need to explore and to forage are considered to be essential elements for a good animal life and therefore necessary in order to justify animal use.

The focus on natural living cannot be recognized only in the public debate. It can also be seen in research on animal welfare. This shows that research also includes aspects such as 'mental state', preference satisfaction and species-specific behaviour. To explain the notion of natural living and species-specific behaviour, Fraser (1999) refers to the Aristotelian concept of *telos* that has been elaborated by Rollin (2004): '… animals have natures of their own (*telos*), and interests that flow from these natures…' From this perspective, natural living implies that the animal has to be able to live according to its nature.

The emphasis on natural living has implications for the relevance and applicability of the notion of longevity as a welfare issue. The question is no longer whether animals suffer as the result of an early death, but to what extent an early death frustrates the aspect of natural living and the ability to express species-specific behaviour. An example of such a view is Nussbaum's capabilities approach. She argues that humans and animals have certain species-specific goals and that these beings should be able to develop these specific abilities and skills in order to flourish and to live according to their *telos*. In this view, longevity clearly becomes a welfare issue, because the possibility of developing species-specific abilities refers not only to functioning and feeling in the present – as it is traditionally perceived in the views on functioning and feeling – but also refers to a certain lifespan. Welfare in terms of flourishing and living a life that is in line with the *telos* of that subject implies a welfare approach that assesses welfare over time and adopts an integral perspective. The time factor emphasizes that welfare should be measured and assessed over a longer period rather than at single points in time. Consequently, lifespan is a relevant welfare criterion. The integral perspective, in which more things are of importance than just the functioning and feelings of animals in the present, implies that the relevance of longevity is not restricted to animals' capacity to be aware of their own future, because flourishing is worth striving for in itself. Because this implies a certain lifespan, which enables animals to develop their full behavioural repertoire, longevity is a constitutive element of animal welfare.

This view supports the moral intuition that killing animals raises moral questions. From this perspective, there is a clear need to justify premature culling of animals, because an animal's lifespan is awarded more value than in the views on welfare that focus on health and feeling well. In those restricted views, culling healthy or curable animals is not seen as a welfare issue for the individual animal. In the views that relate to natural living, culling is seen as having a negative impact on animal welfare. Longevity then serves as an independent moral argument for animal welfare. This means that prematurely culling animals for farm-technical reasons becomes an animal welfare issue, irrespective of whether or not the animal endures direct physical or mental discomfort. This role of longevity in animal welfare requires the inclusion in animal welfare assessments of the integral lifespan and the development and capacities of the individual animals.

10.5 The role of longevity in animal welfare assessment in practice

In order to illustrate the practical implications of claims relating to the importance of longevity, the above analysis will be applied to examples in dairy farming and sow breeding.

10.5.1 Modern animal farming and resulting dilemmas

As described in the introduction, over the last few decades animal production has intensified and specialized considerably, leading to the use of specialized production lines and increased production levels. The developments and related welfare issues will be described briefly for both dairy farming and pig breeding.

Dairy farming

Dairy farming has the reputation of being an extensive way of farming. However, modern farms are no longer all that extensive. The average number of dairy cows per farm is about 85 milking cows (CRV, 2013), which represents a large increase during the last few decades. And for coming years a strong increase is expected in the number of cows kept per farm due to the abolition of the milk quota. Milk production per cow has risen substantially as well. On average, a dairy cow produces more than 8,200 kg of milk per year (CRV, 2013), which represents a doubling over a 60-year period. In north-western Europe, the most common housing system is the cubicle housing system, allowing a large number of cows to be handled in a labour-efficient way. In a cubicle housing system dairy cows can lie and rest in the cubicles and move around in the walking alleys. Housing and management raise a number of health and welfare issues. Some of these problems can lead to the decision to cull an animal (and replace it with another animal). On average, a cow is replaced when she is about six years of age (CRV, 2013) and there is a tendency in farm management to try and increase this. FAWC states that it is reasonable to strive for an average age of eight years (FAWC, 2009). The most important production diseases that lead to premature culling are mastitis, infertility, and foot disorders and lameness. The rate of replacement is about 30% per year (CRV, 2013). The impact on the welfare of the cow varies according to the disease and its severity, but it also depends on how comfortable the animal's surroundings are (which depends on housing and management).

On a conventional dairy farm, many cows have a foot disorder that is present but is not easy to detect (subclinical foot disorder). About a quarter of these cows become lame (clinical foot disorder) after a while (Somers *et al.*, 2003). Measures such as improving the lying places or the walking surface could help prevent new cases or support healing and reduce the pain of existing cases. As foot disorders and lameness are the most important welfare issue in dairy farming, and often lead to premature culling (directly or indirectly) (EFSA, 2009), this production disease will serve as an example.

Sow breeding

Swine production is known for its intensive character. For example, in the Netherlands the number of sows per specialized breeding farm was around 500 animals in 2012, which represents a doubling over ten years. Moreover, as in many agricultural businesses, the number of farms is decreasing while the number of animals per farm is increasing (LEI, 2011). In the European Union, it has not been permitted for sows to be housed individually since 2013. Sows may only be housed individually during the days around insemination and in the farrowing pen (EC, 1998; Varkensbesluit, 2012). The number of piglets produced during a sow's life is 66, with 28 piglets per sow per year on average (Hoste, 2013). The average age that a sow reaches has declined over recent decades. In the 1970s sows were kept for five to eight years[2]. Currently, sows are replaced at an age of about 2.5 years (Klein Swormink, 2011). On a farm, about 42% of sows are replaced annually (Gezondheidsdienst voor Dieren, personal communications). There are various reasons for replacing sows: reproduction problems (30%), locomotion problems (11%), death (11%), euthanasia (mainly due to locomotion, bad udder or other characteristics, 5%), selection/insufficient production (20%), age (14%) and diverse (9%) (Gezondheidsdienst voor Dieren, personal communications). These numbers show that breeding sows can have different health problems that lead to culling. The decision to cull a sow is mostly based on economic and practical considerations. A breeding sow with fertility problems or a smaller litter size will be culled at the end of the gestation period, usually from the fifth or sixth litter onwards. As reproductive problems are the most important reason for culling breeding sows, this will serve as an example relating to sow breeding.

---

[2] http://tinyurl.com/ph4xpoq.

10.5.2 The role of longevity in animal welfare: practical implications

Animal welfare without the notion of longevity

If the view is adopted that longevity is not relevant for animal welfare discussions, then premature culling is not problematic. This is often considered to be the prevailing view among scientist and farmers. As a consequence, no justification is needed for culling animals prematurely. Premature culling solely for farm-technical reasons is not assessed as a welfare problem, as long as animals are killed without pain or stress. From this point of view, investment in order to increase lifespan is not seen as a real welfare improvement. In practice this means that culling animals solely for farm-technical reasons is not a moral issue. For instance, when breeding sows produce fewer piglets (because of reduced litter size or more still-born piglets due to the increasing age of the sow) it is not a problem to cull the sow prematurely. From this perspective, lifespan does not play an independent role in the decision whether or not to cull an animal. Similarly, in dairy farming, culling healthy dairy cows to replace them with young stock with higher production potential is also not considered to be a welfare problem according to this prevailing view.

If a cow has a clinical foot disorder and is consequently lame, it is even considered positive for animal welfare to cull the animal prematurely. This is because the premature culling prevents such an animal from suffering pain and thus negative feelings. Investment in measures that reduce the negative impact of disorders and help to prevent them (such as improved lying places and walking alleys) is not of added value as long as it does not increase productivity substantially. According to this view, it does not matter for animal welfare whether an animal is culled to eliminate a disease or whether more effort is made to prevent or cure diseases.

Justification of culling based on purely economic and farm-technical arguments is accepted by those with views on animal welfare in which longevity serves as an indicator or precondition for animal welfare.

Longevity is an independent moral argument for animal welfare

If longevity is accepted as a constitutive element of animal welfare, premature culling needs more justification, because it is then seen as an act that affects animal welfare. Culling for farm-technical reasons, such as replacing dairy cows by superior young stock, is an animal welfare issue according to this view. The view that longevity is an independent moral argument for animal welfare also includes the opinion that animals should be able to express species-specific behaviour and to flourish. Certain farming practices should be changed in order to do justice to these aspects of animal welfare. Another important implication is that prevention of health and welfare problems should be accorded higher priority in order to prevent premature culling.

Practices to improve the ability of animals to perform species-specific behaviour and to flourish relate to adjustments in housing and management that will improve animals' living conditions. At the same time, these measures should protect the animals from disease, which should prevent premature culling. Such measures do not always contribute (directly) to efficient production, and thus economic viability. One example is that of improving comfort for breeding sows by providing more bedding material and more space, which is less efficient and requires greater management efforts to prevent higher mortality in new-born piglets (e.g. Van Nieuwamerongen *et al.*, 2014). However, in some cases such measures can be interesting from an economic point of view as well. In dairy farming, for example, more measures should be taken to prevent foot disorders and to improve the chance of recovery. This can be achieved by providing better walking and lying surfaces, which will improve comfort for the cow. This will have a positive effect on animal welfare (i.e. feeling and natural living) and can be achieved with measures that can be cost-effective at the same time (Bruijnis *et al.*, 2013a). It is also important to note that premature culling is still a moral wrong, even when the animal itself does not experience discomfort. The common practice in pig breeding of culling a sow after the fifth litter due to decreased reproductive performance, for instance, also becomes more problematic. Increasing the lifespan in this case does not require measures to improve animals' ability to perform important behaviours or to flourish, but only to keep them alive for longer, despite decreased reproductive performance (or to find a way to improve reproductive performance at a greater age). Such changes might be difficult to achieve, as this example negatively correlates with the economic performance of a farm, which is of course key for its viability. It is important to seek approaches that change practices in such a way that the advantage of a longer life can be translated into added value for the economic performance of a farm.

Accepting longevity as an animal welfare issue would bring other dilemmas to the fore. If animals have a longer productive life, new moral dilemmas will require moral assessment. In dairy farming, for example, fewer replacement animals would be needed if dairy cows had a longer productive life. At the same time, for each lactation a calf is born. This means that a dairy cow would produce more calves in her lifetime, but fewer calves would be needed. In the current situation, most female calves replace the dairy cows and most male calves and the females that are not needed for replacement become veal calves or are culled (without being slaughtered for meat). In the new situation, more female calves would be culled. This example reveals a new moral dilemma, which might be partly solved by prolonging the lactation period (resulting in fewer calves per dairy cow). A more thorough analysis of such 'collateral damage' is given in Aerts and De Tavernier (2016).

10.6 Conclusions

The interpretation of the concept of animal welfare is important for decisions on whether or not to cull animals (Haynes, 2016; Rollin, 2016) but also for the measures that should be taken to prevent premature culling. My proposal is to accept the view on animal welfare according to which longevity is accepted as an independent moral argument. Acceptance of this view substantiates the intuition that premature culling of animals is a moral wrong.

The interests of animals are given more weight when the view is adopted that longevity serves as an independent moral argument for animal welfare. In order to respect this view, some common practices in animal farming will become the subject of debate, as illustrated in the two cases. In these debates, interests other than those of the animals also need to be included. Such interests include the economic viability of farms and labour conditions for farmers. It is a challenge to respond to such dilemmas in a balanced way.

References

Aerts, S. and De Tavernier, J., 2016. Killing animals as a matter of collateral damage. In: Meijboom, F.L.B and Stassen, E.N. (eds.) The end of animal life: a start for ethical debate. Wageningen Academic Publishers, Wageningen, the Netherlands, pp. 169-185.

Balcombe, J., 2009. Animal pleasure and its moral significance. Applied Animal Behaviour Science 118: 208-216.

Bradley, B., 2009. Well-being and death. Oxford University Press Inc., New York, NY, USA, 224 pp.

Broom, D.M., 2007. Quality of life means welfare: how is it related to other concepts and assessed? Animal Welfare 16: 45-53.

Bruijnis, M.R.N., Hogeveen, H. and Stassen, E.N., 2013a. Measures to improve dairy cow foot health: consequences for farmer income and dairy cow welfare. Animal 7: 167-175.

Bruijnis, M.R.N., Meijboom, F.L.B. and Stassen, E.N., 2013b. Longevity as an animal welfare issue applied to the case of foot disorders in dairy cattle. Journal of Agricultural & Environmental Ethics 26: 191-205.

Centraal Bureau voor de Statistiek (CBS), 2014. Available at: http://statline.cbs.nl.

Clayton, N.S., Bussey, T.J. and Dickinson, A., 2003. Can animals recall the past and plan for the future? Nature reviews. Neuroscience 4: 685-691.

Cohen, N.E., Brom, F. and Stassen, E.N., 2009. Fundamental moral attitudes to animals and their role in judgment: an empirical model to describe fundamental moral attitudes to animals and their role in judgment on the culling of healthy animals during an animal disease epidemic. Journal of Agricultural and Environmental Ethics 22: 341-359.

Cohen, N.E., Van Asseldonk, M.A.P.M. and Stassen, E.N., 2007. Social-ethical issues concerning the control strategy of animal diseases in the European Union: a survey. Agriculture and Human Values 24: 499-510.

Coöperatie Rundveeverbetering (CRV), 2013. CRV jaarstatistieken 2013. Available at: http://tinyurl.com/pjfmo7p.

De Cock Buning, T., Pompe, V., Hopster, H. and De Brauw, C., 2012. Denken over dieren – dier en ding, zegen en zorg. Inventarisatie van publiekswaarden en verwachtingen van prakijken in 2011. Athena Instituut, Vrije Universiteit Amsterdam, Amsterdam, the Netherlands, 82 pp.

DeGrazia, D., 1996. Taking animals seriously: mental life and moral status. Press Syndicate of the University of Cambridge, Cambridge, UK, 316 pp.

De Greef, K., Stafleu, F. and De Lauwere, C., 2006. A simple value-distinction approach aids transparency in farm animal welfare debate. Journal of Agricultural and Environmental Ethics 19: 57-66.

European Commission (EC), 1998. Council Directive 98/58/EEC. Official Journal of the European Union L 221: 23-27. Available at: http://tinyurl.com/ol6chk5.

European Commission (EC), 2009. Treaty of Lisbon. Official Journal of the European Union C 306. Available at: http://tinyurl.com/ntlh4wb.

European Food Safety Authority (EFSA), 2009. Scientific opinion on the effects of farming systems on dairy cow welfare and disease. Scientific opinion of the Panel on Animal Health and Animal Welfare. EFSA Journal 1143: 1-38.

Farm Animal Welfare Council (FAWC), 2009. Opinion on the welfare of the dairy cow. FAWC, London, UK, 16 pp.

Food and Agriculture Organization of the United Nations (FAO), 2012. Global trends and future challenges for the work of the organization. FAO, Rome, Italy, 89 pp. Available at: http://tinyurl.com/pvb3pzo.

Fraser, D., 1999. Animal ethics and animal welfare science: bridging the two cultures. Applied Animal Behaviour Science 65: 171-189.

Fraser, D., Weary, D.M., Pajor, E.A. and Milligan, B.N., 1997. A scientific conception of animal welfare that reflects ethical concerns. Animal Welfare 6, 187-205.

Haynes, R.P., 2008. Animal welfare: competing conceptions and their ethical complications. Springer, Dordrecht, the Netherlands, 162 pp.

Haynes, R.P., 2016. Killing as a welfare issue. In: Meijboom, F.L.B and Stassen, E.N. (eds.) The end of animal life: a start for ethical debate. Wageningen Academic Publishers, Wageningen, the Netherlands, pp. 41-48.

Hoste, R., 2013. Productiekosten van varkens. Resultaten van InterPIG over 2011. LEI Wageningen UR, Den Haag, the Netherlands, 43 pp.

Klein Swormink, B., 2011. Langere levensduur zonder in te leveren. Available at: http://tinyurl.com/qajttuq.

Landbouw Economisch Instituut (LEI), 2011. BINternet: bedrijfsresultaten en inkomens van land- en tuinbouwbedrijven. Available at: http://www.agrimatie.nl/binternet.aspx.

Lund, V. and Olsson, I., 2006. Animal agriculture: symbiosis, culture, or ethical conflict? Journal of Agricultural and Environmental Ethics 19: 47-56.

McMahan, J., 2002. The ethics of killing, problems at the margins of life. Oxford University Press, New York, NY, USA, 556 pp.

Nussbaum, M.C., 2006. Frontiers of justice: disability, nationality, species membership. The Belknap Press of Harvard University Press, Cambridge, UK, 512 pp.

Regan, T., 1983. The case for animal rights. University of California Press, Berkeley, LA, USA, 425 pp.

Rollin, B.E., 2004. Annual meeting keynote address: animal agriculture and emerging social ethics for animals. Journal of Animal Science 82: 955-964.

Rollin, B.E., 2006. Euthanasia and quality of life. JAVMA 228: 1014-1016.

Rollin, B.E., 2016. Death, *telos* and euthanasia. In: Meijboom, F.L.B and Stassen, E.N. (eds.) The end of animal life: a start for ethical debate. Wageningen Academic Publishers, Wageningen, the Netherlands, pp. 49-59.

Rutgers, L.J.E., Swabe, J.M. and Noordhuizen-Stassen, E.N., 2003. Het doden van gehouden dieren, ja, mits... of nee, tenzij? Universiteit Utrecht, the Netherlands, 118 pp

Sapontzis, S.F., 1987. Morals, reason, and animals. Temple University Press, Philadelphia, PA, USA, 328 pp.

Schmidt, K., 2011. Concepts of animal welfare in relation to positions in animal ethics. Acta Biotheoretica 59: 153-171.

Singer, P., 1993. Practical ethics. Cambridge University Press, New York, NY, USA, 396 pp.

Somers, J.G.C.J., Frankena, K., Noordhuizen-Stassen, E.N. and Metz, J.H.M., 2003. Prevalence of claw disorders in Dutch dairy cows exposed to several floor systems. Journal of Dairy Science 86: 2082-2093.

Taylor, P.W., 1986. The attitude of respect for nature. Respect for nature: a theory of environmental ethics. Princeton University Press, Princeton, NJ, USA, pp. 59-86.

Van Nieuwamerongen, S.E., Bolhuis, J.E., Van der Peet-Schwering, C.M.C. and Soede, N.M., 2014. A review of sow and piglet behaviour and performance in group housing systems for lactating sows. Animal 8: 448-460.

Varkensbesluit, 2012. Available at: http://wetten.overheid.nl/BWBR0006806/geldigheidsdatum_26-09-2012/afdrukken.

Yeates, J., 2009. Death is a welfare issue. Journal of Agricultural and Environmental Ethics 23: 229-241.

Part III.

Killing in different practices of animal use

11. Killing animals as a matter of collateral damage

S. Aerts[1,2], and J. De Tavernier[2]*
[1]Odisee University College, Hospitaalstraat 23, 9100 Sint-Niklaas, Belgium; [2]Ethics@ Arenberg, KU Leuven, Sint-Michielsstraat 4, P.O. Box 3101, 3000 Leuven, Belgium; stef.@odisee.be

Abstract

Not only meat producing animals are killed in agriculture. Also in the dairy and egg industry enormous numbers of animals are killed, although their deaths are not strictly necessary to produce milk or eggs. These deaths are a side effect of current economic realities and are considered unavoidable collateral damage. We will discuss other cases such as culling during disease control, and euthanasia of aged sports animals and animals in shelters. Other examples are fishing discards, dying animals in nature reserves, culled hobby animals. All these examples are characterised by a systematic killing of animals. These animals are not or no longer needed and the killing appears as an unavoidable side effect of a particular production type or husbandry system. It is therefore distinct from accidental killings or killing for meat production. A second important distinctive criterion is the feeling of meaninglessness or disproportionality connected to these practices. Killing as collateral damage is a non-issue from an animal rights ethics viewpoint because from this perspective any kind of killing is considered unethical. On the other hand, in utilitarian and hybrid anthropocentric-zoocentric approaches that integrate proportionality in their reasoning, it is considered a moral problem. In many cases, an analysis of the different (moral) costs and benefits is difficult because killing these animals is considered to be a side effect of other activities rather than an activity with its own value. There seem to be two alternatives: either the benefits are divided between the intended killings and the collateral killings, or only the secondary goal is allocated to the collateral killings. In either case, the ratio is heavily skewed to the negative side. Except in extreme anthropocentric theories killing animals as collateral damage seems at least problematic, if not extremely problematic.

Keywords: animal disease control, culling, instrumentalisation, proportionality, intentionality

11.1 Introduction

As humans, we usually decide whether the direct or indirect killing of animals is allowable in order to sustain our lives. Inspired by Jeremy Bentham, the founding father of 'classic' utilitarianism, in his influential book 'Animal liberation' Peter Singer

*Franck L.B. Meijboom and Elsbeth N. Stassen (ed.) **The end of animal life: a start for ethical debate***
DOI 10.3920/978-90-8686-808-7_11, © Wageningen Academic Publishers 2016

(1975) explicitly states the killing of animals is not wrong *per se*. The core of Singer's argument is that one should not only balance the good and bad consequences for humans. In explaining his position, Singer mainly focuses on pain and suffering (which he uses almost synonymously). If it is considered wrong to hurt humans, he argues, it is wrong to inflict pain or suffering to all beings that are able to suffer (which includes most animals). In other words, Singer works with the principle of 'least suffering' from a non-speciecist position. This is why his utilitarian ethics could also be called 'pathocentrism' (De Tavernier and Aerts, 2003).

Singer's arguments are relevant in all contexts in which animals are a part of the human sphere. Next to the use of experimental animals in cosmetic and medical research, it is especially relevant to animal agriculture. Even though his perspective could *a priori* allow for the rearing and killing of animals for food, Singer favours a vegetarian position, all consequences taken into consideration, vegetarianism is the right option as it minimises suffering of sentient beings. For him, not eating animal meat is the lesser evil and it is the only acceptable outcome of a reasonable impartial moral cost-benefit analysis.

Another line of argument is put forward by Tom Regan (1983) in his book 'The case for animal rights', who holds that animals have (moral) rights (e.g. the right to life) and that an ethically right action consists of non-violating these rights. Regan's deontological approach implies that the only justified option is to stop animal production immediately and to become vegan, because the use of milk, eggs, wool, leather, honey etc. is no longer a legitimate praxis.

Regan's reaction to Singer is that morally good attitudes towards animals need more than 'equal consideration of equal interests'. He further elaborates on his thoughts in 'Defending Animal Rights' (Regan, 2001), in which he differentiates between the defence of animal welfare, the defence of animal rights and opposing cruelty to animals. Opposing cruelty is always opposition against unnecessary suffering, which implies that there are occasions in which it is acceptable (or even proportional) to hurt animals (e.g. in case of animal experiments in which inflicting pain for a greater good is acceptable). According to Regan the philosophy of animal welfare, which goes further than opposition against cruelty, is also insufficient (and even inappropriate) because it also allows for infringements on the integrity of animals in case of a proportional reason. Indeed, only defending animal welfare (as in Singer's utilitarian philosophy) still allows animals to be raised for human ends. Regan argues that animal production *in se* is morally wrong, not because animals are hurt, frustrated, stressed or killed, but because it violates their moral rights. The animals' right to be respected, according to Regan, directly leads to the position that they should never be treated as a means to an end.

It should be noted that the above discussion indicates that in a utilitarian or consequentialist ethic, inflicting pain or suffering on an animal is acceptable when there is a proportional reason. In a deontological approach of ethics, proportionality is not used.

11.2 Killing animals as a matter of collateral damage

In our contemporary society, vegetarians, and certainly vegans, are a small minority. At this point in time we can only conclude that a large majority in our society – knowingly or unknowingly – deems the killing of animals for food production morally acceptable. As a consequence, an extremely large number of animals are reared every year in order to accommodate our appetite for animal products. Obviously, meat production can only be done by killing animals; something meat eaters consider to be a proportional reason (or their deontological right), while vegetarians do not.

Less obvious is that egg and milk production also results in many animals killed each year. Laying hens and dairy cows are killed after a (short) productive life. In addition, and more importantly, almost all male animals of these breeds are killed long before they reach adulthood. Indeed, raising male animals of highly specialised breeds in egg and milk production has increasingly become an economic impossibility. In the current production context, driven by efficiency, killing these animals is unavoidable. A simple matter of 'collateral damage'.

When do we classify the killing of animals as collateral damage? To us, it seems there are two important elements. The first element is intentionality. Collateral damage in the common sense is 'during a war, the unintentional deaths and injuries of people who are not soldiers, and damage that is caused to their homes, hospitals, schools, etc.' (Cambridge Dictionary, 2014). This (un)intentionality can be translated directly to killing of animals. That would mean that in this chapter we would only have to include situations in which animals are killed accidentally, e.g. animals that are killed by farmers during routine field activities. Although this situation is not devoid of ethical issues, it is not what we would like to call 'killing as collateral damage'. Here we discuss only situations that include intentional killing as an active intervention. A death that does not contribute to the goal one is trying to achieve, but nevertheless seems unavoidable or cannot be avoided. The production and the killing of these animals is a side effect of other activities rather than an activity on its own. The meaning of 'unintentional' in this chapter is therefore more like 'originally unplanned' or 'unintended', but with a distinct pattern or systematic approach.

The second, and equally important criterion, is meaninglessness or absence of proportionality. We would include only those activities that invoke a sense of meaninglessness, or even alienation. This criterion is the reason why we do include the killing of day-old male chicks, but not the rats that are used as live food for carnivorous pets. In an ethical vocabulary, this would often translate in a lack of proportionality. Killing as collateral damage therefore does not include only those killings that are 'useless' *sensu stricto*, but also those where the use seems disproportionately insignificant (some of those chicks are indeed used as food for pets or zoo animals). When there is a use, it is often more an afterthought than a goal.

In this chapter we analyse which ethical and societal questions and problems arise in different cases of 'collateral killing' of animals, and we look for appropriate ways to handle these particular end of life discussions.

11.3 Cases of collateral killing

In many different areas of human activities, animals are being killed as a 'side effect' or as 'collateral damage'. In this chapter we will present three general cases that cover a broad range of settings and moral challenges. We will discuss the killing of day-old chicks as it is the most prominent case, but also because this is a case in which many of the typical characteristics of collateral killing are most prominent. Secondly, we will look at killing animals during disease control. This has been a matter of vivid discussion over the last decade, and recent events in Europe have made it highly relevant again. The last case we will discuss is the killing of animals that no longer 'function' in their intended role as pets are sports animals.

A case that we will not discuss here is that of the unintentional – but to a certain extent planned, predictable, and preventable – 'death-by-neglect' of wild animals in nature reserves. Gamborg *et al.* (2010) have already discussed the difficulties one encounters when trying to identify the ethically right treatment of such animals. This brought Swart and Keulartz (2011) to introduce a contextual element into the debate in order to be able to identify the amount of care that is appropriate. Although this case can be interpreted in the context of collateral killing of animals, it is discussed already at length in a separate chapter in this book (Gremmen, 2016).

As said, egg production is one of the most prominent cases in which animals are killed as a side effect of production techniques. The development of specialised layer breeds has led to the situation in which it is no longer economically viable to raise the male animals. No farmer would want to feed these males several times the amount of feed a broiler chicken would need during a much longer time. Instead, the male chicks are killed before they are delivered to the farms. A similar situation arises in the dairy sector. The male calves of the specialised Holstein-Friesian dairy herd, which is half of all the dairy calves born annually, are not suited for beef production. Sometimes they are used for veal production, but in many cases, these calves are killed shortly after their birth. In 2010 the British organic agriculture organisation, opposing the typical veal production techniques, committed to phasing out the culling practice (Soil Association, 2014). In order to limit the amount of technical details, we will focus our discussion on the chick case; *mutatis mutandis* the analysis will translate to the dairy case.

Disease control is another practice in which a great number of animals are killed as a side effect. Killing healthy animals as part of a prevention strategy is considered to be an epidemiologically successful decision in particular cases. It does explain, to some extent, why in 2001 at least 250,000 animals were culled in the Netherlands because 26 farms became infected with foot and mouth disease (FMD) (EC, 2001). During the avian influenza (AI) outbreak of 2003, there were 255 infected farms in the Netherlands and eight in Belgium. This caused 30.3 million and 4.1 million birds to be culled, respectively (FASFC, 2004; LNV, 2004; Stegeman *et al.*, 2004).

The final case we will explore is the euthanasia of animals such as aging sports animals (mainly horses) and animals in shelters. Both are cases where animals have lost their intended 'function'. The goal of animal shelters is to save animals from undesirable situations (stray animals, abused animals, unwanted animals, etc.) and try to provide a new home for them. This is not always easy, or even possible. We have seen considerable public outcry over the last few years since statistics from People for the Ethical Treatment of Animals animal shelters in the USA have been disclosed. Public data released via websites such as www.petakillsanimals.com indicate that some shelters euthanize almost all of the animals they receive. We will not discuss this individual case any further, but it is a clear indication that in animal shelters quite some 'collateral' killing is done.

11.4 Killing of one day-old chickens

As we have argued before (Aerts *et al.*, 2009), killing day-old chickens is a highly polemic practice for a number of reasons. To the general public, it is not only little-known but therefore also very 'surprising' when visually confronted with this killing practice on small and furry animals. A good illustration of this is the public's reaction to the gassing of chicks in Jamie Oliver's 'Fowl dinners'. Nevertheless, it is an almost century old practice (dated from the 1920's, when specialised layer breeds became common). On a yearly basis around 4 billion of these 'useless' male chicks are produced (280 million in Europe, 230 million in the USA), which is a conservative estimation as there are 5.7 billion laying hens (Evans, 2007). Right after the chicks are released from the hatcher, the male chicks are selected manually and are killed immediately by electrocution, maceration, or CO_2 gassing; only the latter two are allowed in the EU.

It is clear that the killing of such a high number of animals cannot be considered an ethically neutral practice or as business as usual. This kind of collateral damage merits an in-depth discussion on the why and how.

A Dutch preliminary analysis of the case (Woelders *et al.*, 2007) identified six elements in an ethical framework with regard to the killing of day-old chicks: animal welfare, wastage, death as an exponent, human health, animal integrity, and naturalness. Although we consider five of these elements as relevant to our analysis, we would like to approach these in this chapter through two broader items: age at killing and instrumentalisation. The latter would then include not only the idea that the animal is becoming a 'by-product' of egg production, but also that its integrity is violated, and that such a system is unnatural. To these categories, we add sustainability issues.

We will not discuss the animal welfare issues raised by the killing methods used on day-old chicks, however relevant. The question whether killing animals is a welfare issue *per se*, is discussed elsewhere in this book, and the welfare issues themselves, do not fall within the scope of this book.

A very prominent aspect of killing male chicks is the age at which the animals are killed. Together with the techniques used (gas suffocation and shredding), the very young age at which the animals are killed is probably the main reason why killing of day-old chicks raises so much debate when it is brought to the attention of the general public. Few other animal production types involve the killing of animals immediately after birth, except for the male offspring of dairy cows and dairy goats in certain parts of the world.

As Woelders *et al.* (2007) already stated, it is difficult to coherently argue why killing day-old chicks would be an ethical problem, but not the slaughter of broiler chickens at an age of five weeks. Also in veal, lamb, and pork production animals are killed long before adulthood. In fact, in all animal production systems animals are killed at relatively young ages, and very few – if any – will reach their 'natural' maximum age.

The 'age at killing' argument is certainly not something that is unique to the case of day-old chicks, but that does not render it more or less important. The fact that it may well be the most pronounced example, on the other hand, is important. We could say it is a situation that shows *par excellence* that animals are bred, used, and killed for human goals. This brings us to the next ethical element in our analysis: instrumentalisation.

The sense that the killing of day-old male chicks is considered as the ultimate example of the boundless instrumentalisation of animals in modern animal production 'industry' seems to be the key point of the opposition against the killing of male chicks at early age. Killing animals so shortly after their birth because they are of no further use than (in the optimal cases) to serve as food for other animals seems to be a grave infringement of a male chicken's *telos*.

That male chickens are seen only as a by-product of egg production and their killing as unavoidable, i.e. collateral damage, is something that can hardly be contested. This does not necessarily mean that it is intrinsically wrong. Whether the integrity of a chicken is more harmed by being killed within the first 48 hours of life, or by being kept for 72 weeks for egg production and then being killed, is doubtful. Again, we seem to encounter a gradual difference from other – less offending – practices when we approach this from an integrity-based analysis.

The same line of thought is applicable to the 'naturalness' argument mentioned by Woelders *et al.* (2007). The 'age at killing', the 'naturalness', and the integrity based 'instrumentalisation' argument are all prone to the accusation of being arbitrary statements. Indeed, it is difficult to identify which practice is right and which one is wrong when these practices differ only gradually over a very wide and continuous scale. Just as it is difficult to identify an acceptable killing age that does not violate a chicken's *telos*, it is unclear what would be the 'natural' condition that would oppose the 'unnatural' killing of day-old chicks. It cannot be that in nature chicks do not die within their first 48 hours of life, because quite some do. However, it is difficult to find another condition that fits well. Probably because there is none. Woelders *et al.* (2007) already said: 'As soon as naturalness no longer exists as a factual (biological) arrangement ... 'naturalness' can no longer act as an independent source for limits to [makeability]' (translated footnote in Woelders *et al.*, 2007: 11). When we start making our own definitions of the 'natural' state of affairs in order to prove a point, we lose all firm ground. Not only the 'age at killing' and the 'naturalness', but also the integrity-based 'instrumentalisation' argument seem to be flawed in this respect when used in a way that treats the killing of extremely young animals differently from the other practices.

On a more practical level, one could also argue that there are sustainability issues connected to the killing of day-old chicks. Without analysing these in detail, it is clear that not only the social component of sustainability is at stake here, but also the economic and ecological components could be discussed. Killing such a number of chickens is an economic loss, since one has spent a lot of energy and effort in producing them. Indeed, significant amounts of inputs are used to produce eggs and chicks that are not useful for the goal they are produced for (egg production). Half of the parent stock, and its associated feed and energy, and half of the hatchery capacity, and its associated energy consumption, are 'useless'.

Are there alternatives to the current practices? And if there are, are they ethically acceptable? In 2003, De Tavernier *et al.* have already evaluated the use of sexed sperm as an alternative to the killing of the male offspring of dairy cattle. This biotechnological fix does not appear to have any adverse consequences for animals as it does not inflict any pain, it has no influence on species-specific behaviour, it does not violate the interests of animals (being free of hunger, thirst, fear, ...), nor does it violate the integrity of the animals. On the other hand, there is a distinct odour of increasing instrumentalisation to this alternative.

In the egg production sector, we see the same lock-in. Moving from post-hatch sex determination to pre-hatch has significant ethical advantages, certainly if it can be done before the full maturation of the central nervous system. Leenstra *et al.* (2010) have looked to the societal approval of different alternatives, and they found that killing embryos instead of day-old chicks was not considered as a real alternative.

The ideal situation would be the case in which the eggs could still be used. There are, for example, already prototypes of methods that assess sex discriminating factors in the late embryonic or foetal stages. One may question, as did the participants in Leenstra *et al.* (2010), whether there is a fundamental difference between the killing of a grown foetus and the killing of a day-old chick. It may only 'hide' the practice for the public, but that is an esthetical difference at best. The ethical issues remain. If there are welfare problems during killing, these will also be applicable to these un-hatched chicks. Neither will it fundamentally solve the age issue, nor the instrumentalisation question (especially when it is done late in the incubation); it only shifts it to a still younger age.

If these techniques can be developed further so that we are able to determine the sex of the embryo at day ten or even earlier, this would deal with an important number of ethical issues. For example, the sex would be known before the nervous system is fully developed, and most welfare issues could be eliminated. If the sex could be determined before incubation, effects that are even more positive can be expected.

There are also some possibilities to use genetic modification (GM) techniques to avoid having to kill male chicks. One could for example introduce a lethal gene variant on the male Z chromosome. In birds, male animals are homozygotes for the sex chromosome (ZZ), and therefore in these GM animals only the heterozygote (ZW) female offspring will develop. It is clear that this GM solution needs to be embedded in a broader debate on GM animals. At this time, it still seems a no-go-zone, although 'changing the chicken by genetic modification to facilitate sexing of freshly laid eggs' scored relatively high in Leenstra *et al.* (2010).

Nevertheless, it seems that alternatives that result in less (or no) male offspring – GM or not – are prone to the instrumentalisation critique. Preventing the conception or the birth of half of the offspring seems not less instrumentalising than killing them.

Summarizing the day-old chick case, it seems that moving from post-hatch sex determination to pre-hatch has significant advantages, not only societally, but also ethically if it can be done before the full maturation of the central nervous system. Unfortunately, this does not avoid all ethical questions, as is shown in the paragraphs on age and instrumentalisation of animals.

11.5 Killing as disease control

In case of necessity, so-called preventive measures are taken to lower the likelihood of disease outbreaks, and to prevent an outbreak to become an epidemic. These measures are based on – or at least comply with – the relevant EU legislation, such as, for example the EU Directive 2005/94/EC (EC, 2005). The classic scenario, used in Belgium during the AI outbreak of 2003, is a 'stamping out' (i.e. culling) of production animals in a 3 km zone, of hobby animals in a 1 km zone, and a 10 days national standstill (prohibition of transport of relevant animals, and animal products). In the 2014 outbreaks, in Germany, the Netherlands and the UK, a similar approach was taken. In the Netherlands, a national 72 hour standstill was complimented with extensive screening and a 30 days standstill in the 10 km zone, and culling of the infected farms and farms in close contact. In Germany all poultry within a 3 km radius has been culled (admittedly, this was a small number), and in the UK a similar 3 km protection zone and 10 km surveillance zone was defined. In the event of an animal disease outbreak, such measures are used to decrease the spreading of the agent or vector. In ethics, this is called the minimum evil (*minus malum*).

One would expect such a strategy to be economically beneficial as well, as lowering the number of diseased animals is a cost effective measure. It is exactly this viewpoint that has made this classic scenario the preferred strategy, and not (truly) preventive vaccination. Indeed, the EU decided in the early 1990's to install a non-vaccination policy for FMD, classical swine fever (CSF) and AI for economic reasons, suggesting that this particular prevention strategy may be more cost effective than vaccination because some countries do not allow import of meat from vaccinated animals (EC, 1992, 2005, 2010).

Decisions about stamping-out strategies (sometimes on a massive scale, e.g. the epidemic of the FMD in 2000 in the UK that took a cull of around 4 million animals) are taken by public authorities. Since public health is one of the concerns of the common good (*bonum commune*) for which governments are supposed to take responsibility (beneficence), they feel urged to eradicate healthy animals in order to prevent further disease spread and safeguard the economically important export of animals. At one hand, governments are requested by their citizens to improve public health protection and to stimulate economic activities and trade. At the other hand, with every new outbreak the public outcry grew since 2000. Since then, culling animals on such a scale have been heavily criticised. Right now, it is no longer a politically acceptable option in Western European democracies. Since there is no longer a consensus on the strategy to be used with regard to future disease outbreaks, governments are developing an increased interest in management techniques that minimise the harm caused by stamping-out strategies. For instance, they invest in developing early warning systems of disease detection, which is often reflected in the urge to introduce into this field the ethical principle of a duty-to-know, overriding the farmer's autonomy. Governments think that they can perform better if informed at early stage about a disease outbreak. Although it has probably not originated from this kind of ethical argument, such a duty-to-know approach has become already an internationally recognised principle in animal disease management. This new approach to disease outbreak management can be seen in the difference between the strategies used in 2003 and 2014 that we discussed earlier in this paragraph.

Are there other alternatives? Many veterinarians think that vaccination could help in order to prevent or in case of an outbreak to limit the size of an epidemic. The larger the epidemic, the higher the need to eradicate a high number of animals. Minimalizing any kind of delay in detecting a notifiable disease, which is mentioned on the yearly updated list of the World Organisation for Animal Health (OIE, 2014), is very important. Once the outbreak is manifest, epidemiologists know that vaccination will have little or no effect. It is simply too late. On the other hand, it is also true that vaccination threatens livestock production since many countries refuse imports from countries that vaccinate because it is very difficult to find out if animals are infected or immunised.

We will not discuss here whether killing during disease control is inherently more likely to induce pain or suffering compared to conventional slaughter. There may be reasons to believe so (Aerts, 2006; Aerts *et al.*, 2006), but practices are too varied to be discussed within the limits of this case. Next to the pathocentric analysis, there are other ethical principles that are definitely at stake here, at different levels. We see that instrumentalisation (related also to a lack of proportionality) is certainly a major point at the societal level, and so is sustainability. If we could avoid killing so many animals without being able to use them in the food chain (through vaccination or otherwise), we see advantages regarding proportionality, economic and ecologic benefits. At the individual level (farmers) this is a case in which autonomy is clearly under stress, maybe more so than in other examples of collateral killing.

11.6 Early killing of 'less functional' animals

For severely ill animals, there are legal standards in veterinary practice entitling veterinarians to end that suffering and provide a 'good death', regardless of the purposes for which the animals are used. The amelioration of suffering is considered a sufficient reason for killing, i.e. euthanasia. However, what should we do with non-suffering healthy animals? What should be decided if animals are no longer capable of performing any of the functions that humans desire, such as producing a high amount of milk or a high number of eggs, or providing companionship? There are many other illustrations to give, e.g. what to do with horses that are no longer able to participate in races? Once high production is over and top fitness goals are no longer reached, the majority of these animals are killed or euthanized. Is questioning these practices reasonable? Do we really have to send them to retirement farms?

In many areas of the world, especially the Anglo-Saxon countries, eating horsemeat is being looked upon with a great deal of horror. The continuing discussion on the re-opening of horse slaughter facilities in the USA is summarised in the 2014 article in 'USA Today' (Massey, 2014). In other countries, such as Belgium, it is a common practice, albeit less so than other meat consumption. This means that a large number of horses are excluded from the food chain after the end of their 'career' as a sports horse or recreative horse. Belgian legislation now permits owners to exclude a horse permanently from the human food chain immediately after birth.

In this regard, it is worth mentioning that tens of thousands of pets, mainly cats and dogs, are brought to shelters every year. Often, owners are hoping that their once loved animals could be re-homed. Nevertheless, the facts are such that – in most countries – for many reasons the great majority of these animals do not have the luck of finding a new home. Killing animals is therefore a common practice in shelters. As a sign that this practice is very controversial, so-called 'no-kill shelters' have been established, giving expression to the idea that the animal's life is itself of value. Since animals are living beings, being alive is according to some 'a strong reason for letting that animal go on living' (Sandøe and Christiansen, 2007). In this viewpoint ending the life of an animal prematurely is surely a bad thing.

What a good animal life constitutes is not a simple question (Appleby and Sandøe, 2002). In decisions about euthanasia 'not only the quality of the animal's continued life is at stake, but also the moral loss involved in ending the life of the animal' (Sandøe and Christiansen, 2007). Since killing humans, even in cases where life is miserable, exception made for consent in specific conditions, is seen as murder, why would we evaluate the lives of animals differently? Moreover, even if there is a distinction to make between the moral status of an animal life and the moral status of a human life, we may not take early killing of animals for granted, simply because their functionality is diminishing (for the animal itself, or in relation to human expectations). From an ethical viewpoint, not only quality of life but also quantity of life is important. Balancing both seems to be difficult.

Another element does play an important role in this discussion. Are animals replaceable? Well-known is Singer's utilitarian argument that it is allowed to kill animals for meat if they are killed painlessly after having a pleasant life and are replaced. In his maximising animal welfare viewpoint, replacement is a key issue. The idea of replacement will be accepted much easier for animals in livestock production than for companion animals. The specific human-animal bond gives impetus to the idea that companion animals are irreplaceable, because, as individuals, they share part of the life history of particular individual people over time. Especially bonds of friendship encourage a view on animals as irreplaceable. The key question here is if saving animal lives outside this personal bond of friendship is important or not. Utilitarians will defend the view that saving animal lives should be considered rather as a personal preference, not as a moral imperative, while deontologists like Regan will defend the opposite.

Again we see the same underlying principles emerge from our discussion. Nevertheless, instrumentalisation is apparent in a slightly different form. There is little difference between the killing of pets in shelters (these have lost their function) and the killing of chicks (that have no function) as long as we disregard the human-animal bond that is present with companion animals. Including this in the discussion reveals the precarity of the replaceability argument, that is implicitly assumed with production animals.

11.7 General discussion

What can be concluded from these preliminary assessments? Is there nothing more to these matters than a more extreme (but not intrinsically different) incarnation of the issues we encounter in any assessment of animal production? If there is more, then what is it that makes these cases different? If there is nothing more, then we are looking at proportionality issues. Therefore, we need to ask if and why killing day-old chicks is problematic or more so than any other animal production activities. Alternatively, is this a 'marginal case' that informs us about the more general issue? And what would it then tell? Does it tell us that nothing is wrong with animal production or that everything is wrong with it?

It is not necessary to discuss killing as collateral damage in animal rights ethics, as it clearly holds all animal use (and therefore certainly killing) as morally wrong. On the other hand, in non-extreme anthropocentric utilitarianism, or hybrid forms that integrate proportionality, it is an issue. In many cases, a cost-benefit analysis is difficult to make because the production and the killing of these animals is a side effect of other activities rather than an activity on its own. When there is a use, it is often more an afterthought than a goal.

Either the benefits are divided between the intended killings and the collateral killings, or only the secondary goal is allocated to the latter. In either case, the ratio is heavily skewed to the negative side.

The different examples described above all have their own ethical issues that seem very different at a first glance. In order to focus on the elements specific for killing as collateral damage, we will not linger on the issues of pain and suffering that are associated with the different examples. The day-old chick case clearly demonstrates the systematic nature of the collateral killing. It also illustrates that many of the issues associated with this case are issues of instrumentalisation and sustainability. A similar image emerges from the disease control discussion. Killing less animals (or none at all) to fight disease outbreaks seems to deliver advantages regarding proportionality, economic and ecologic benefits. From the disease eradication case and the shelter euthanasia case we distil an additional element within the instrumentalisation argument; that of the implicit assumption of replaceability of animal lives. It emerges most noticeably in circumstances in which the human-animal bond is individual and therefore very clear, but it is present in all cases.

In short, we believe there are three general themes to be distinguished that are applicable to all cases where animals are killed as collateral damage. These are: (1) instrumentalisation; (2) sustainability; and (3) ethically irrelevant aesthetics.

Inherent to our definition of 'killing as collateral damage' is that (the death of) the animal is a by-product of another activity. Animals such as day-old chicks are produced as by-products and then killed. In most (if not all) examples there are clear indications of a violation of the (physical) integrity of the animals concerned, or that they are prevented from fulfilling their *telos*. In our examples, we have mentioned the feeling of unnaturalness that surrounds some of the activities that lead to collateral killing. We have not always found anything (more) unnatural in these practices than in conventional husbandry. Violation of physical integrity, not fulfilling their telos, and unnaturalness are all linked to a more general concept: instrumentalisation. Some of our examples are even considered the 'ultimate example of the boundless instrumentalisation of animals' in modern animal production 'industry'.

If we look at the three basic components of sustainability (society, economy and ecology), it is clear that the current rate of collateral killing is unsustainable. The outcry that has followed the public disclosure of some of these practices ('Fowl dinners', petakillsanimals.com, etc.) is a perfect indication that there is a serious problem with the social acceptance of the killing of animals as collateral damage. This is, of course, quite contradictory with the fact that most people are (in)directly involved in the activities that lead to this killing. Going into this apparent contradiction is beyond the scope of this chapter, but some of the reasoning, showing that this is not always real contradictory behaviour, can be found elsewhere (Aerts, 2013; Aerts and Lips, 2010). Most probably any progress made towards less collateral killing would also result in significant economic and ecologic gains. There are obvious direct gains by eliminating costs made to kill and process these animals, but also – and arguably vastly more significant – indirect gains by avoiding unnecessary costs made in the activities that led to this side effect. If no day-old chicks or male dairy calves are produced, only half of the number of parent stock would be needed, half of the incubation and hatching capacity, etc. Similar analyses can be made for the other cases, even the euthanasia of animals in shelters. All this current 'overcapacity' readily translates in economic and ecologic costs. A great move towards more sustainable animal production (or whatever other activity concerned) would be made by a diminished number of animals killed as collateral damage.

To us, one of the most striking facts in this discussion is that there is in fact not much difference with conventional practices. Billions of animals are killed on a routine basis, every year. Animals die, and are killed, long before they reach their maximum age. Animals are culled when they reach the end of their productive lives. *In se*, there are few differences between conventional husbandry and collateral killing. These features may be more apparent, more extreme, and more visually disturbing, but these are esthetical rather than ethical issues. In short, these elements only seem to separate the killing of animals as collateral damage from other, more conventional practices.

11.8 Conclusions

Except in the extreme anthropocentric theories, killing animals as collateral damage seems at least problematic, if not extremely problematic.

However, based on our discussion in the previous paragraphs it seems that these practices are not inherently different from many other practices that are under much less (public) debate. It is likely a matter of proportionality, a gradual difference in which the cost-benefit analysis of collateral killings is more negative than that of conventional practices.

Our ethical analysis indicates that the killing of animals as collateral damage does not result in problems other than those inherent to our animal use as a whole. The fundamental opposition raised against the killing and the instrumentalisation of animals is inherent to animal production (conventional or other), and is certainly not limited to the examples presented here. They may be more apparent, maybe more extreme and visually more disturbing, but these are barely ethically relevant. It is difficult to condemn (strongly) the collateral killings as wrong, without questioning keeping and producing animals in general.

This is a slippery slope most people would not be prepared to go. Nonetheless, it is a (higher-order) debate that merits attention. In the end, it may well prove that the collateral killing case is a marginal case that informs us about the ethical status of our relation to animals as a whole.

References

Aerts, S., 2006. Practice-oriented ethical models to bridge animal production, ethics and society. PhD dissertation. KU Leuven, Faculty of Bioscience Engineering, Leuven, Belgium, 200 pp.

Aerts, S., 2013. The consumer does not exist: overcoming the citizen/consumer paradox by shifting focus. In: Röcklinsberg, H. and Sandin, P. (eds.) The ethics of consumption: the citizen, the market, and the law. Wageningen Academic Publishers, Wageningen, the Netherlands, pp. 172-176.

Aerts, S. and Lips, D. 2010. Dierenwelzijn en consumptie. In: De Tavernier, J., Lips, D. and Aerts, S. (eds.) Dier en welzijn. LannooCampus, Leuven, Belgium, pp. 83-97.

Aerts, S., Boonen, R., Bruggeman, V., De Tavernier, J. and Decuypere, E., 2009. Culling of day-old chicks: opening the debates of Moria? In: Millar, K., Hobson West, P. and Nerlich, B. (eds.) Ethical futures: bioscience and food horizons. Wageningen Academic Publishers, Wageningen, the Netherlands, pp. 117-122.

Aerts, S., Evers, J. and Lips, D., 2006. Development of an animal disease intervention matrix (ADIM). In: Kaiser, M. and Lien, M. (eds.) Ethics and politics of food. Wageningen Academic Publishers, Wageningen, the Netherlands, pp. 233-238.

Appleby, M.C. and Sandøe, P., 2002. Philosophical debate on the nature of well-being: implications for animal welfare. Animal Welfare 11: 283-294.

Cambridge Dictionary, 2014. Meaning of 'collateral damage' in the English Dictionary. Available at http://dictionary.cambridge.org/dictionary/english/collateral-damage.

De Tavernier, J. and Aerts, S., 2003. Het doden van dieren: ethische aspecten. In: Koolmees, P.A., Swabe, J.M. and Rutgers, L.J.E. (eds.) Het doden van dieren. Maatschappelijke en ethische aspecten. Wageningen Academic Publishers, Wageningen, the Netherlands, pp. 35-45.

De Tavernier, J., Lips, D., Delezie, E., Aerts, S., Van Outryve, J., Spencer, S. and Decuypere, E., 2003. Ethical considerations on the usage of sexed sperm in stock breeding. In: Jolivet, E. (ed.). Ethics as a dimension of agrifood policy. Proceedings of the 4th Congress of the European Society for Agricultural and Food Ethics. INRA, Toulouse, France, pp. 174-175.

European Commission (EC), 1991. Council Directive 91/685/EEC of 11 December 1991 amending Directive 80/217/EEC introducing community measures for the control of classical swine-fever. Official Journal of the European Union L377: 1-14.

European Commission (EC), 1992. Council Directive 92/40/EEC of 19 May 1992 introducing community measures for the control of avian influenza. Official Journal of the European Union L167: 1-16.

European Commission (EC), 2001. Final report of the international conference on control and prevention of foot and mouth disease. December 12-13, 2001. EU Presidency, Brussels, Belgium.

European Commission (EC), 2005. Council Directive 2005/94/EC of 20 December 2005 on community measures for the control of avian influenza and repealing Directive 92/40/EEC. Official Journal of the European Union L10: 16-65.

European Commission (EC), 2010. Council Directive 90/423/EEC of 26 June 1990 amending Directive 85/511/EEC introducing community measures for the control of foot-and-mouth disease, Directive 64/432/EEC on animal health problems affecting intra-community trade in bovine animals and swine and Directive 72/462/EEC on health and veterinary inspection problems upon importation of bovine animals and swine and fresh meat or meat products from third countries. Official Journal of the European Union L224: 13-18.

Evans, T., 2007. Growth slows in global demand for poultry products. Poultry International 46: 12-16.

Federal Agency for the Safety of the Food Chain (FASFC), 2004. Activity report 2003. FASFC, Brussels, Belgium.

Gamborg, C., Gremmen, B., Christiansen, S.B. and Sandøe, P., 2010. De-Domestication: ethics at the intersection of landscape restoration and animal welfare. Environmental Values 19: 57-78.

Gremmen, B., 2016. Will wild make a moral difference? In: Meijboom, F.L.B and Stassen, E.N. (eds.) The end of animal life: a start for ethical debate. Wageningen Academic Publishers, Wageningen, the Netherlands, pp. 251-263.

Leenstra, F., Munnichs, G., Beekman, V., Van den Heuvel-Vromans, E., Aramyan, L. and Woelders, H., 2010. Killing day-old chicks? Public opinion regarding potential alternatives. Animal Welfare 20: 37-45.

Massey, B., 2014. Obama signs budget measure that withholds money for required federal inspections of slaughter process. Available at http://tinyurl.com/opoa33t/.

Ministerie van Landbouw, Natuurbeheer en Voedselkwaliteit (LNV), 2004. Avian influenza in the Netherlands. Outbreak 2003 – regionalisation. Presentation at the SPS enhanced meeting on Article 6 (regionalisation). January 30-31, Geneva, Switzerland. Available at: http://tinyurl.com/ngaauww.

Regan, T., 1983. The case for animal rights. University of California Press, Berkeley, CA, USA, 448 pp.

Regan, T., 2001. Defending animal rights. University of Illinois Press, Urbana, IL, USA, 179 pp.

Sandøe, P. and Christiansen, S.B., 2007. The value of animal life: how should we balance quality against quantity? Animal Welfare 16: 109-115.

Singer, P., 1975. Animal liberation. A new ethics for our treatment of animals. Random House, New York, NY, USA.

Soil Association, 2014. Dairy cattle. Available at http://tinyurl.com/ozq4fo7.

Stegeman, A., Bouma, A., Elbers, A.R.W., De Jong, M.C.M., Nodelijk, G., De Klerk, F., Koch, G and Van Boven, M., 2004. Avian influenza A virus (H7N7) epidemic in the Netherlands in 2003: course of the epidemic and effectiveness of control measures. Journal of Infectious Diseases 190: 2088-2095.

Swart, J.A.A. and Keulartz, J., 2011. Wild animals in our backyard. A contextual approach to the intrinsic value of animals. Acta Biotheoretica 59: 185-200.

Woelders, H., Brom, F.W.A and Hopster, H., 2007. Alternatieven voor doding ééndagskuikens: technologische perspectieven en ethische consequenties. Rapport 44. Wageningen Universiteit & Research Centrum, Animal Sciences Group, Lelystad, the Netherlands, 29 pp.

World Organisation for Animal Health (OIE), 2014. OIE-listed diseases, infections and infestations in force in 2014. Available at http://tinyurl.com/po46y2b.

12. Killing animals as a necessary evil? The case of animal research

N.H. Franco[1,2] and I.A.S. Olsson [1,2*]

[1]Instituto de Investigação e Inovação em Saúde, Universidade do Porto, Rua Alfredo Allen 208, 4200-135 Porto, Portugal; [2]IBMC-Instituto de Biologia Molecular e Celular, Universidade do Porto, Rua Alfredo Allen 208, 4200-135 Porto, Portugal; olsson@ibmc.up.pt

Abstract

This chapter addresses the question of killing animals in research, primarily from a moral perspective, but also taking into account some of the practical and scientific considerations with moral consequences in this context. We start by exploring in which situations animals are killed in research and whether these are always inevitable, analysing re-use and re-homing of animals as potential alternatives. We then discuss for whom – and under what circumstances – killing matters, considering situations where there may be a conflict between the wish to avoid killing and that to avoid suffering, and further take human-animal interactions into account. We argue that, although there are relevant practical, scientific and ethical arguments favouring the euthanasia of animals in most research contexts, there is a potential for rehabilitating more animals than is currently the practice.

Keywords: laboratory animals, euthanasia, reuse, rehabilitation, rehoming

12.1 Introduction

The use of animals in the life sciences seems to be accepted by most people provided that it allows advancing biomedical knowledge, that such advances cannot be achieved using non-animal methods and that animal suffering is kept to a minimum (Aldhous *et al.*, 1999; Crettaz von Roten, 2012). Nevertheless, animal experimentation remains a controversial issue. While the use of animals in general and the welfare of these animals have been subject of wide debate, within the discussion of animal experimentation the moral implications of *killing* animals, have not been given as much attention.

Typically, animals used in research are euthanized at the end of the experiments. The most recent statistics (EC, 2013) regarding the scientific use of animals in the European Union point to a total of 11.5 million vertebrates used in 2011 in all fields of basic and applied biomedical science, as well as in education and training, in both the public and the private sector. Considering the size of EU27 population – 502.5 million people in 2011 (Eurostat, 2012) – this gives a ratio of roughly 2.3 vertebrates (mostly

Franck L.B. Meijboom and Elsbeth N. Stassen (ed.) The end of animal life: a start for ethical debate
DOI 10.3920/978-90-8686-808-7_12, © Wageningen Academic Publishers 2016

rodents and fish) used per 100 EU citizens every year. This makes the annual number of animals used in Europe for all scientific and educational purposes but a very small fraction of that of those – mostly cows, pig, sheep, poultry and fish – killed for food in the EU daily (for statistics see Eurostat, 2008, 2009; 2011). Of course, the fact that the overwhelming majority of people (including those concerned about animal welfare) seem to approve of the killing of animals for food, is not a reason to dismiss ethical concerns over the killing of animals in research. First, the use of animals may be more readily and easily replaced in food production than in biomedical research (Cohen, 1986), at least as regards the nutritional value. Maybe more important, using the majority view as moral guidance is questionable to say the least. Also, whereas meat production without killing seems inconceivable, research and experiments[1] may not necessarily require the curtailing of animals' lives. Furthermore, the fact that the number of animals killed in experiments pales in comparison to the vast numbers killed in common human activities outside the laboratories does not remove our moral responsibilities towards these animals.

Starting from the assumption that at least some animal research is relevant, ethically acceptable and presently not replaceable, some harm to animals in research may be perceived as a 'necessary evil', in particular in face of the moral importance of advancing biomedical knowledge for the benefit of humans and non-humans alike. However, it should nevertheless be reflected upon in which circumstances it may – or may not – be either 'necessary' or 'evil' to kill animals in the context of animal research. In this chapter, we discuss whether killing is inevitable, or morally problematic, as well as to whom this killing matters.

12.2 Killing animals in research – is it always inevitable?

In order to answer the question of whether killing is inevitable, it becomes necessary to understand in which situations, and for what reasons, animals are killed in research. The majority of cases fall into three main categories, namely (1) when the research as such requires that animals are killed, (2) when killing is (or is considered to be) in the animals' best interest to prevent further suffering, and (3) when killing results from a contingency – of financial, logistic, technical or even cultural nature – secondary to the scientific process, *per se*. The first two cases present a scientifically and ethically more convincing argument for the euthanasia of animals, as the unavoidability of killing in these cases is taken as a starting premise. However, alternative approaches to some experimental procedures that typically require killing may be considered.

---

[1] This chapter refers specifically to the killing of animals in a biomedical context. Although equally important, other contexts where animal research takes place, such as research on farm animals, present distinct ethical, technical and social implications which we do not address here.

The killing of laboratory animals is often elicited by the need to collect appropriately sized biological samples in smaller species, like rodents or fish, where for example the entire volume of blood may be required for analysis. When this is combined with the need to collect samples at different time points, an animal – or a group of animals – for each time point must be killed. In some cases, necropsy is needed to assess the internal alterations caused by disease. When using larger animal species, however, repeated sampling can be accomplished without having to kill different animals at several time points, with the added advantage that the quality of the experimental outcome may be improved by minimizing the effect of inter-individual variability, since the same animal is used as its own control (Bloomsmith *et al.*, 2006; Dell *et al.*, 2002). Necropsy can also sometimes be replaced by use of imaging technologies to follow disease progress in the living animal, while avoiding having to kill animals for this purpose. One example is the use of imaging technologies to follow the progress of developing infections such as in tuberculosis research (Andreu *et al.*, 2012; Davis *et al.*, 2009a,b) as well as the growth of tumours in both rodents and non-human primates. This allows reducing considerably the number of animals otherwise needed for *post-mortem* analysis at different time points.

Laboratory animals may also be killed in order to prevent them from unnecessary suffering. A paramount example is the use of 'humane endpoints', generally understood as the euthanasia of research animals when their health and welfare reach a previously defined threshold level of pain or suffering. Humane endpoints are particularly important in studies on progressive diseases, as they prevent that animals reach advanced stages and subsequently die from the disease or associated conditions (e.g. starvation or dehydration due to inability to reach the food hopper; or attack by less affected cage mates). They may also be applied in response to unexpected welfare problems requiring emergent intervention (resulting from injury, procedural errors or sudden aggravation of clinical signs). In such cases, when animals would otherwise suffer and this suffering cannot be avoided in any other way, the early killing of research animals is generally considered to be the best practice, as well as often legally required (Morton, 1999).

Laboratory animals are however sometimes killed for what seem to be rather trivial reasons. One derives from a tendency to use only animals of one sex. This is sometimes females, in order to avoid aggression-related problems with group-housed males (Van Loo *et al.*, 2003, 2004) but more often males are preferred (Beery and Zucker, 2011; Wald and Wu, 2010). In any case, preference for a given sex can lead to the culling of animals of the other sex. Also, from a scientific perspective, using both male and female animals in research is valuable since it allows detecting possible sex differences, which with appropriate experimental design can be achieved without using additional animals. Such routine culling of healthy animals may also result from insufficient planning of experiments, since one reason animal facility staff and commercial

12.3 Is it morally 'wrong' to kill animals in research?

In the previous section we focused on practical issues as regards finding alternatives to the early curtailing of animals' lives in research. It remains to be discussed whether, or to which degree, killing an animal is, or may become, a moral issue.

Historically, the Western tradition of thinking does not consider the killing of non-humans morally problematic in itself. The Judaeo-Christian religious moral tradition held that while one must abstain from cruelty, killing animals was not in itself morally problematic, as animals lacked an 'immortal soul' (Franco, 2013). In the secular anthropocentrism that would follow from the seventeenth century on, cruelty towards animals would continue to be considered morally condemnable but, as Immanuel Kant would state, those 'who use living animals for their experiments, certainly act cruelly, although their aim is praiseworthy, and they can justify their cruelty, since animals must be regarded as man's instruments' (Kant, 1997). Even Jeremy Bentham, founder of utilitarian moral philosophy[2] would not state animal research to be unethical, provided the experiment had 'a determinate object, beneficial to mankind, accompanied with a fair prospect of the accomplishment of it', thus acknowledging humans had certain precedence over other animals (Boralevi, 1984).

The present mainstream approach to animal use in research is predominantly utilitarian in nature, and in general greater attention is given to preventing the suffering of animals than to avoiding their killing. This may sometimes have quite far-reaching consequences. Under current European legislation regulating animal use in experiments, killing laboratory animals is not even – by definition – considered to be a procedure, if no other prior interventions are carried out (EC, 2010). This is consistent with the predominant *welfarist* view of good practice in research with animals, under which the painless killing of laboratory animals poses no ethical problem, or at least not from a welfare point a view, as non-existence inheritably implies absence of negative experiences. Following the welfarist approach, conducting experiments under terminal anaesthesia is seen as an ethically preferable approach, as animals are not aware of any aversive stimuli, and are spared the distress associated with recovery from anaesthesia and other interventions.

However, the assumption that death in itself is not a moral problem is not as clear-cut for every researcher, and for every species. When nineteenth-century physician George Hoggan said he would be 'inclined to look upon anaesthetics as the greatest curse to vivisectible animals', he alluded to the fact that anaesthesia allowed researchers to use great numbers of animals without further moral quandaries. More recently,

---

[2] In proposing sentience as the primary criterion for defining to whom should be given moral consideration, Bentham built the philosophical framework within which Peter Singer operated in his seminal work *Animal Liberation* (Singer, 2002).

James Yeates challenged the mainstream idea that the swift and painless killing of animals is not a welfare issue, defending instead that there are at least some instances in which killing animals poses a welfare problem. He bases his argument by following the rationale that it is in an animals' interest to not only avoid negative feelings but also experience positive ones, from where it follows that as long as the animal would be expected to have a life worth living, death deprives the animal of the positive feelings it would otherwise experience in its lifetime (McMahan, 2008).

Animal research presents a number of situations where there is an apparent conflict between avoiding killing and preventing suffering. While reduction and refinement[3] may often go hand-in-hand, there are several instances in which these principles conflict with each other. If one were to be governed by the 'badness of killing' argument, one would as far as possible avoid taking the lives of animals, thus giving precedence to reduction. However, it is more widely recognized that refinement should be prioritized if reduction efforts result in a significant burden for each individual animal that would otherwise be avoided or minimized by using more animals (De Boo *et al.*, 2005; Franco and Olsson, 2014; Hansen *et al.*, 1999). In fact, by requiring that the application of reduction should take 'into account individual animal welfare in relation to minimizing pain, suffering, distress or lasting harm', current texts partly address this conflict and establish that avoidance of killing should not come at the expense of animal suffering.

Nevertheless, this precedence is not consensual. Having the refinement/reduction conflict in mind, we presented a markedly dichotomic reduction/refinement dilemma to participants in 11 laboratory animal science courses held in Portugal between 2008 and 2011 (n=235). Asked whether they would rather use twenty mice on a painful experiment (but with no permanent physical damage) or use the same mouse in repeated measures (assuming this would not impact the validity of the study), 49% would rather use the same mouse, while 51% preferred to divide the burden by 20 mice. A proportion of researchers would, however, reconsider and reuse the same animal if it were a dog (31%) or a non-human primate (38%). But even if the proportions are different for different species, the overall picture is still that these animal researchers divide into a group who give preference to avoiding suffering and another group preferring to avoid killing.[4]

---

[3] Replacement (of animal experiments with alternative approaches), reduction (of animal numbers) and refinement (of experimental procedures to reduce animal pain, distress and suffering) constitute the 3Rs. First presented in 1959 (Russell and Burch, 1959), this is now a widely accepted governing principle for animal research.

[4] A more comprehensive view of scientists' attitudes to animal research can be found in Franco and Olsson (2014), which reports a study conducted on a large part of this sample. For further discussion on the conflict between longevity, value and quality of life in animal research and other contexts, see Franco *et al.* (2014).

12.4 To whom does death matter?

It seems obvious that the main stakeholders in this issue are the animals themselves; after all it is their lives that are ended or allowed to continue. However, trying to approach the question of killing from an animal-centred perspective means confronting a number of philosophical questions that, for us humans, are troublesome to answer (see Frey, 2011). Understanding that one day one's life will end and one will cease to exist is part of the human experience and development. In some way, death seems potentially more harmful to an individual who has this awareness, because they can be harmed also by worrying about the timing, conditions and consequences of death in the future (McMahan, 2002). Our very limited understanding of what death means to an animal is a strong limitation when discussing the ethics of killing. Equally important and difficult to answer is the question of what is a life worth living and when death is preferable to going on living. These questions are developed in more detail in other chapters in this book and we will not attempt to answer them here. However, as the next example will illustrate, even taking a more anthropocentric perspective on animal killing does not completely free us from questions about what death means to animals.

Philosophers such as Bernard Rollin (2016) have suggested that humans have established social contracts with domesticated animal species, under which both species have benefited across time, such as in traditional farming (Rollin, 2011). This concept may also be applied to the collaborative relationship between researchers and laboratory animals (Iliff, 2002). It is reasonable to say that, for instance, rodents used in non-invasive experiments and under good husbandry are likely to be, at least in some aspects, better off than their wild counterparts[5], as they are protected from threats to their wellbeing such as weather conditions, hypothermia, dehydration, starvation, natural predators or pest control and have their basic needs catered for by humans. In the usual interpretation of this hypothetical social contract, humans get to decide when and how to kill animals. If accepting the argument that it is better to have lived a short, but 'happy' life than not to have lived at all, when should we consider that animals do not live a 'life worth living' and we hence fail to our part of the contract? Is the use of animals lives less justified in these cases? If so, and if rehabilitation is possible, does allowing an animal to retire after its scientific purpose help fulfil our duty of fairness to the animals?

Important as they are, the animals are not the only stakeholders in this discussion. The use of animals in research typically requires instrumentalizing them to some degree, but animals are certainly not – and are not perceived as – inert instruments of inquiry.

---

[5] For instance, even if provided with shelter and *ad libitum* food and water in a semi-natural enclosure (Calhoun, 1963) the average life span of wild rats (*Rattus norvegicus*) is about half of that of laboratory rats of the same species (Altun *et al.*, 2007).

Instead, they are living beings that can establish significant relationships with humans. This animal-human bond is particularly strong between animals and caregivers who, rather than associating animals with a given procedure or test, see these sporadic interventions as only a small fraction of the animals' lives and of their time spent with them. This bidirectional interaction leads staff members to develop an appreciation for the value of the animals' lives, which in turn makes their death matter morally to them (Bayne, 2002; Chang and Hart, 2002; Cressey, 2011; Holmberg, 2008; Iliff, 2002). Typically, laboratory animal caretakers are also often those who have to kill the animals trusted to their care. The contradiction between these tasks can elicit what A. Arluke (1994) has coined as the 'caring-killing paradox', characterized by a sentiment of grief, guilt and moral stress, even in situations when euthanasia is performed to prevent debilitated animals of further suffering. Nevertheless, getting to know the animals and establish a bond with them may be important for allowing caretakers to cope with stress and improve animal health, well-being, and the quality of research, in spite of the emotional cost borne by the animal facility personnel.

People working in animal facilities feel particularly downhearted about killing healthy animals for what they may perceive as convenience, as when an experiment is coming to an end, animals are being excluded from an experiment (e.g. for failing to perform a designated task in a behavioural study); or during routine culling of surplus animals in animal facilities, the latter rendering killing a mere management technique. In these situations, and especially when animals have not served any scientific purpose, aside the grief that may arise, killing will also be perceived as 'wrong' to students, veterinarians and caregivers, in particular when a viable adoptive home, or other alternative, is available (Birke *et al.*, 2007; Carbone, 1997; Chang and Hart, 2002).[6]

From the human perspective, it seems as if all animals are not equal when it comes to killing. This is evident in the most recent EU legislation regulating animal use, in which it says that 'animals such as dogs and cats should be allowed to be rehomed in families as there is a high level of public concern as to the fate of such animals' (EC, 2010: Recital 26). A similar species preference is also shared by animal facility personnel (Chang and Hart, 2002; Cressey, 2011) and was also evident among researchers participating in laboratory animal science training. Indeed, many were open to allow several animal species – in particular non-human primates and companion animal species – to be given for adoption or taken to a sanctuary if rehabilitation were possible (Figure 12.1).

12.5 Discussion

In this chapter, we have discussed whether the killing of laboratory animals is always inevitable, whether it is morally problematic and to whom such killing matters. As we

---

[6] Researchers, on the other hand, have been described as having more of an ambivalent view (Birke *et al.*, 2007; Lynch, 1988).

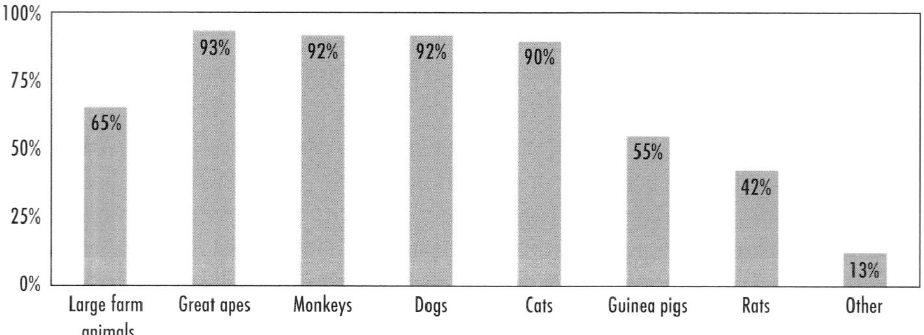

Figure 12.1. Data from a survey conducted with participants in 11 laboratory animal science courses held in Portugal between 2009 and 2011 (n=193, courses held in four different universities and following FELASA guidelines for either category B or C).

demonstrated in the first section, it is often unavoidable to end the lives of research animals, either because the research itself requires this or because the animals would otherwise suffer unacceptable pain and distress. However, animals are sometimes killed when they could have been rehabilitated or even when they are healthy and have not been used in research. In these situations, re-homing into sanctuaries or family homes is a viable option, which is presently not extensively used. In the second section, we discussed the moral implications of killing animals. Even though in the mainstream view on laboratory animal ethics preventing suffering seems to be more important than avoiding killing, killing nevertheless remains an ethical issue. This is also strongly suggested by the emotional reaction of humans working with animals and having to decide over, and execute, their killing, as discussed in the third section of the paper.

The humans directly or indirectly involved in the killing of research animals seem to differentiate between different animal species, considering it more problematic to kill cats, dogs and non-human primates than rodents and rabbits. This differential attitude is not limited to the killing of animals but applies to animals in research in general, and it is consistent with the socio-zoological scale presented by Arluke and Sanders in which animals are ranked according to how much they are valued by humans, with companion animal species and non-human primates topping the list (Arluke and Sanders, 1996). Consistent with this, there is also a greater investment in finding alternatives to killing animals in the case of dogs (usually re-homed into families) and chimpanzees (retired into sanctuaries).

But even for smaller species, such as rodents, the possibility of rehabilitation and rehoming, at least in some circumstances, may not be so far-fetched. It may in fact sometimes be easier to find a home for groups of these animals than for a single dog. From the perspective of the adopter, the small size, easy and affordable maintenance and short longevity means the commitment might be easier to take on. Also, from the animals' perspective, adaptation may be easier for a rodent who will be maintained with its social group than for a dog who needs to adapt from living primarily with dogs to be a single canine member of a human family. One interesting option for rodents is their use as classroom pets in schools (Baumans *et al.*, 2007). This can be a useful resource for teaching values as responsibility and respect for animals as well as to dealing with loss (Huddart and Naherniak, 2005). When well managed and housed in large and species-appropriate habitats, the animals are provided better living conditions than in the animal facility (Fonseca *et al.*, 2011). No matter whether the animals are adopted to families or to schools, it is of course important that the adopters consciously assume the responsibility to care for the animals as long as these live. Rehoming animals to initially friendly carers who lose interest in these animals in short time is neither in the animals' nor in the adopters' best interests.

Several issues, however, may prevent this practice from becoming mainstream. These include animal rehabilitation being labour demanding, costly and time consuming; the difficulty in ensuring that re-homed animals are housed, supervised and handled at least to the same standards found in animal facilities; and the onus of responsibility – as well as legal liability – in case animals are found to be mistreated. Animals may also manifest physical and behavioural abnormalities with a welfare impact, as a result of prolonged captivity or inadaptation to re-homing conditions (Hubrecht, 2002; Kerwin, 2006). Therefore, rehabilitation of most laboratory animals may be unpractical or even unfeasible, thus often making euthanasia the ethically preferable option. However, although re-homing of research animals may presently sound outlandish to most, it nonetheless deserves more serious thought, in particular for the considerable number of surplus animals that do not need rehabilitation nor constitute a risk of disease transmission. After all, if animal research is usually portrayed to the public as only being carried out when no alternatives are available, should this not also hold for killing?

Acknowledgements

The authors wish to thank Dorte Bratbo Sørensen and Helena Rocklinsberg for constructive criticism of a previous version of this chapter.

References

Aldhous, P., Coghlan, A. and Copley, J., 1999 Animal experiments: let the people speak. New Scientist 162: 26-31.

Altun, M., Bergman, E., Edström, E., Johnson, H. and Ulfhake, B., 2007. Behavioral impairments of the aging rat. Physiology & Behavior 92: 911-923.

Andreu, N., Elkington, P.T. and Wiles, S., 2012. Molecular imaging in TB: from the bench to the clinic. In: Cardona, P.J. (ed.) Understanding tuberculosis – global experiences and innovative approaches to the diagnosis. InTech, Rijeka, Croatia, pp. 307-318.

Anonymous, 2006. Breeding of and experiments on animals (control and supervision) amendment rules. 23 October 2006. In: Ministry of Environment and Forests (ed.) Gazette of India, Extraordinary. Part II, Section 3, Sub-section (ii).

Anonymous, 2010. Tierschutzgesetz. (TierSchG). (Animal Protection Act). December 9th 2010, BGBl. I S. 1934. 32 pp. Available at: http://tinyurl.com/nle6qxv.

Arluke, A. and Sanders, C.R., 1996. Regarding animals. Temple University Press, Philadelphia, PA, USA, 256 pp.

Arluke, A., 1994. Managing emotions in an animal shelter. In: Manning, A. and Serpell, J. (eds.) Animals and human society: changing perspectives. Routledge, New York, NY, USA, pp. 145-165.

Baumans, V., Coke, C., Green, J., Moreau, E., Morton, D., Patterson-Kane, E., Reinhardt, A., Reinhardt, V. and Loo, P.V. (eds.), 2007. Making lives easier for animals in research labs. Animal Welfare Institute, Washington, DC, USA.

Bayne, K., 2002. Development of the human-research animal bond and its impact on animal well-being. ILAR Journal 43: 4-9.

Beery, A.K. and Zucker, I., 2011. Sex bias in neuroscience and biomedical research. Neuroscience & Biobehavioral Reviews 35: 565-572.

Birke, L., Arluke, A. and Michael, M., 2007. The division of emotional labor. In: The sacrifice: how scientific experiments transform animals and people. Purdue University Press, West Lafayette, IN, USA, pp. 93-110.

Bloomsmith, M.A., Schapiro, S.J. and Strobert, E.A., 2006. Preparing chimpanzees for laboratory research. ILAR Journal 47: 316-325.

Boralevi, L.C., 1984. Bentham and the oppressed. Walter de Gruyter GmbH, Berlin, Germany.

Broadhead, C.L., Betton, G., Combes, R., Damment, S., Everett, D., Garner, C., Godsafe, Z., Healing, G., Heywood, R., Jennings, M., Lumley, C., Oliver, G., Smith, D., Straughan, D., Topham, J., Wallis, R., Wilson, S. and Buckley, P., 2000. Prospects for reducing and refining the use of dogs in the regulatory toxicity testing of pharmaceuticals. Human & Experimental Toxicology 19: 440-447.

Calhoun, J.B., 1963. The ecology and sociology of the Norway Rat. U.S. Department of Health, Education, and Welfare, Public Health Service, Washington, DC, USA, 288 pp.

Carbone, L., 1997. Adoption of research animals. Animal Welfare Information Center Newsletter 7: no. 3-4.

Chang, F.T. and Hart, L.H., 2002. Human-animal bonds in the laboratory: how animal behavior affects the perspectives of caregivers. ILAR Journal 43: 10-18.

Cohen, C., 1986. The case for the use of animals in biomedical research. New England Journal of Medicine 315: 865-870.

Cressey, D., 2011. Battle scars. Nature 470: 452-453.

Crettaz von Roten, F., 2012. Public perceptions of animal experimentation across Europe. Public Understanding of Science.

Davis, S.L., Be, N.A., Lamichhane, G., Nimmagadda, S., Pomper, M.G., Bishai, W.R. and Jain, S.K., 2009a. Bacterial thymidine kinase as a non-invasive imaging reporter for *Mycobacterium tuberculosis* in live animals. PLoS ONE 4: e6297.

Davis, S.L., Nuermberger, E.L., Um, P.K., Vidal, C., Jedynak, B., Pomper, M.G., Bishai, W.R. and Jain, S.K., 2009b. Noninvasive pulmonary [18F]-2-fluoro-deoxy-D-glucose positron emission tomography correlates with bactericidal activity of tuberculosis drug treatment. Antimicrobial Agents and Chemotherapy 53: 4879-4884.

De Boo, M.J., Rennie, A.E., Buchanan-Smith, H.M. and Hendriksen, C.F.M., 2005. The interplay between replacement, reduction and refinement: considerations where the Three Rs interact. Animal Welfare 14: 327-332.

Dell, R.B., Holleran, S. and Ramakrishnan, R., 2002. Sample size determination. ILAR Journal 43: 207-213.

DiGangi, B.A., Crawford, P.C. and Levy, J.K., 2006. Outcome of cats adopted from a biomedical research program. Journal of Applied Animal Welfare Science 9: 143-163.

Doehring, D. and Erhard, M.H., 2005. Whereabouts of surplus and surviving laboratory animals. ALTEX 22: 7-11.

European Commission (EC), 2010. Directive 2010/63/EU of the European Parliament and of the Council of 22 September 2010 on the protection of animals used for scientific purposes. Official Journal of the European Union L276: 33-79.

European Commission (EC), 2013. Seventh report on the statistics on the number of animals used for experimental and other scientific purposes in the member states of the European Union. European Union, Brussels, Belgique.

Eurostat, 2008. Poultry statistics in the European Union: flock numbers, hatcheries, trade and slaughterings, Data in focus 31/2008. Eurostat, Kirchberg, Luxembourg.

Eurostat, 2009. EU cattle, pigs, sheep and goats: monthly slaughter statistics in 2008. Data in focus 15/2009. Eurostat, Kirchberg, Luxembourg.

Eurostat, 2011. Fishery statistics, agriculture and fishery statistics – main results 2010-11. Eurostat Pocketbooks. Eurostat, Kirchberg, Luxembourg.

Eurostat, 2012. Europe in figures – Eurostat yearbook 2012. Eurostat, Kirchberg, Luxembourg, 119 pp.

Fonseca, M.J., Franco, N.H., Brosseron, F., Tavares, F., Olsson, I.A.S. and Borlido-Santos, J., 2011. Children's attitudes towards animals: evidence from the Rodentia project. Journal of Biological Education 45: 121-128.

Franco, N.H., 2013. Animal experiments in biomedical research: a historical perspective. Animals 3: 238-273.

Franco, N. and Olsson, I., 2014. Scientists and the 3Rs: attitudes to animal use in biomedical research and the effect of mandatory training in laboratory animal science. Laboratory Animals 48: 50-60.

Franco, N.H., Magalhães-Sant'Ana, M. and Olsson, I.A.S., 2014. Welfare and quantity of life. In: Appleby, M., Weary, D. and Sandøe, P. (eds.) Dilemmas in animal welfare. CABI, Wallingford, UK, pp. 46-66.

Frey, R.G., 2011. Utilitarianism and animals. In: Beauchamp, T.L. and Frey, R.G. (eds.) The Oxford handbook of animal ethics. Oxford University Press, New York, NY, USA, pp. 172-197.

Hansen, A.K., Sandøe, P., Svendsen, O., Forsman, B. and Thomson, P., 1999. The need to refine the notion of reduction. In: Hendriksen, C. and Morton, D. (eds.) Humane endpoints in animal experiments for biomedical research. RSM Press, London, UK, pp. 139-144.

Hawkins, P., Morton, D.B., Bevan, R., Heath, K., Kirkwood, J., Pearce, P., Scott, L., Whelan, G. and Webb, A., 2004. Husbandry refinements for rats, mice, dogs and non-human primates used in telemetry procedures. Seventh report of the BVAAWF/FRAME/RSPCA/UFAW Joint Working Group on Refinement, Part B. Laboratory Animals 38: 1-10.

Holmberg, T., 2008. A feeling for the animal: on becoming an experimentalist. Society and Animals 16: 316-335.

Hubrecht, R., 2002. Comfortable quarters for laboratory dogs in research institutions. In: Reinhardt, V. and Reinhardt, A. (eds.) Comfortable quarters for laboratory animals. Animal Welfare Institute, Washington, DC, USA, pp. 56-64.

Huddart, S. and Naherniak, C., 2005. Animals in the classroom. In: Grant, T. and Littlejohn, G. (eds.) Teaching green: the elementary years: hands-on learning in grades K-5. New Society Publishers, Toronto, Canada, pp. 101-107.

Iliff, S.A., 2002. An additional 'R': Remembering the animals. ILAR Journal 43: 38-47.

Jennings, M., Prescott, M.J., Buchanan-Smith, H.M., Gamble, M.R., Gore, M., Hawkins, P., Hubrecht, R., Hudson, S., Keeley, J.R., Morris, K., Morton, D.B., Owen, S., Pearce, P.C., Robb, D., Rumble, R.J., Wolfensohn, S. and Buist, D., 2009. Refinements in husbandry, care and common procedures for non-human primates. Ninth report of the BVAAWF/FRAME/RSPCA/UFAW Joint Working Group on Refinement. Laboratory Animals 43 (Suppl. 1): 1-47.

Kant, I., 1997. Lectures on ethics. In: Schneewind, J.B. and Heath, P. (eds.) The Cambridge edition to the works of Immanuel Kant. Cambridge University Press, Cambridge, UK.

Kelch, T.G., 2011. Globalization and animal law. Comparative law, international law and international trade. Kluwer Law International, Alphen aan de Rijn, the Netherlands, 520 pp.

Kerwin, A.M., 2006. Overcoming the barriers to the retirement of old and new world monkeys from research facilities. Journal of Applied Animal Welfare Science 9: 337-347.

Kramer, K. and Kinter, L.B., 2003. Evaluation and applications of radiotelemetry in small laboratory animals. Physiological Genomics 13: 197-205.

Laboratory Animal Science Association (LASA), 2004. LASA Guidance on the Rehoming of Laboratory Dogs. A report based on a LASA working party and LASA meeting on rehoming laboratory animals. LASA, Hull, UK. Available at: http://tinyurl.com/nbr5juj.

Lynch, M.E., 1988. Sacrifice and the transformation of the animal body into a scientific object: laboratory culture and ritual practice in the neurosciences. Social Studies of Science 18: 265-289.

McMahan, J., 2002. The wrongness of killing and the badness of death, the ethics of killing: killing at the margins of life. Oxford University Press, New York, NY, USA, pp. 189-202.

McMahan, J., 2008. Eating animals the nice way. Daedalus 137: 66-76.

Morton, D.B., 1999. Humane endpoints in animal experimentation for biomedical research. In: Hendriksen, C.F.M. and Morton, D.B. (eds.) Humane endpoints in animals experiments for biomedical research. Royal Society of Medicine Press, London, UK, pp. 5-12.

Morton, D.B., Hawkins, P., Bevan, R., Heath, K., Kirkwood, J., Pearce, P., Scott, L., Whelan, G. and Webb, A., 2003. Refinements in telemetry procedures. Seventh report of the BVA(AWF)/FRAME/RSPCA/UFAW joint working group on refinement, part A. Laboratory Animals 37: 261-299.

Pereira, S. and Tettamanti, M., 2005. Ahimsa and alternatives – the concept of the 4th R. The CPCSEA in India. ALTEX 22: 3-6.

Rollin, B.E., 2011. Animal rights as a mainstream phenomenon. Animals 1: 102-115.

Rollin, 2016. Death, *telos* and euthanasia. In: Meijboom, F.L.B. and Stassen, E.N. (eds.) The end of animal life: a start for ethical debate. Wageningen Academic Publishers, Wageningen, the Netherlands, pp. 49-59.

Russell, W.M.S. and Burch, R.L., 1959. The principles of humane experimental technique. Methuen & Co, London, UK.

Singer, P., 2002. Animal liberation. Ecco, New York, NY, USA.

Stephens, M.L., Conlee, K., Alvino, G. and Rowan, A.N., 2002. Possibilities for refinement and reduction: future improvements within regulatory testing. ILAR Journal 43: S74-S79.

Turner, P.V., Smiler, K.L., Hargaden, M. and Koch, M.A., 2003. Refinements in the care and use of animals in toxicology studies regulation, validation, and progress. Journal of the American Association for Laboratory Animal Science 42: 8-15.

Van Loo, P.L.P., Van de Weerd, H.A., Van Zutphen, L.F.M. and Baumans, V., 2004. Preference for social contact versus environmental enrichment in male laboratory mice. Laboratory Animals 38: 178-188.

Van Loo, P.L.P., Van Zutphen, L.F.M. and Baumans, V., 2003. Male management: coping with aggression problems in male laboratory mice. Laboratory Animals 37: 300-313.

Waitt, C.D., Bushmitz, M. and Honess, P.E., 2010. Designing environments for aged primates. Laboratory Primate Newsletter 49: 5-9.

Wald, C. and Wu, C., 2010. Of mice and women: the bias in animal models. Science 327: 1571-1572.

Weekley, L.B., Guittin, P. and Chamberland, G., 2002. The international symposium on regulatory testing and animal welfare: recommendations on best scientific practices for safety evaluation using nonrodent species. ILAR Journal 43: S118-S122.

Wolfensohn, S., 2010. Euthanasia and other fates for laboratory animals. In: Hubrecht, R. and Kirkwood, J. (eds.) The UFAW handbook on the care and management of laboratory and other research animals. Wiley-Blackwell, Hoboken, NJ, USA, pp. 219-226.

13. Killing of companion animals: to be avoided at all costs?

J. van Herten
Royal Veterinary Association of the Netherlands, P.O. Box 421, 3995 AW Houten, the Netherlands; j.van.herten@knmvd.nl

Abstract

Looking into end of life decisions concerning companion animals roughly two kinds of issues can be identified. On the one hand we sometimes kill companion animals too late causing unnecessary suffering and on the other hand there are situations in which might we kill them too fast, depriving them a natural lifespan and possible future wellbeing. These situations raise moral questions about role and responsibilities of pet owners and veterinarians and about justification of end of life decisions regarding companion animals. We can address these questions by looking at the implications of moral standing of companion animals in modern Western societies. The so-called human-companion animal bond implies a moral obligation to take the interests of our companion animals seriously into account. I will argue that when making decisions about end of life of our companion animals, the interests of the concerned animal will normally outweigh the interests of the owner. An animal's future quality of life is the most important parameter. We therefore have a moral obligation to euthanize animals in case of unbearable and hopeless suffering. Killing healthy companion animals however can only be justified in special circumstances. To help veterinarians in making difficult end of life decisions, we have developed an assessment model. By using this model veterinarians are guided to carefully weigh all the different interests in play and make justified decisions about killing companion animals.

Keywords: companion animal, killing, veterinarian

13.1 Introduction

In contrast to other domesticated animals, many companion animals are considered members of the family (Endenburg, 1991). Consequently, if they get ill their owners often provide them all the veterinary care that is necessary. This is considered responsible pet ownership (AVMA, 2012). However, in situations of severe illness animals sometimes have to undergo intensive treatments, which include serious side effects, painful recovery or long-lasting medication. Because animals cannot give consent, the owner and veterinarian have the responsibility to make decisions on whether it is better to treat or to euthanize the animal. Owners are often willing to do everything that's possible for their beloved animal. This, however, raises the question whether veterinarians always should accommodate to the owners' requests if these

*Franck L.B. Meijboom and Elsbeth N. Stassen (ed.) **The end of animal life: a start for ethical debate***
DOI 10.3920/978-90-8686-808-7_13, © Wageningen Academic Publishers 2016

are technically feasible? Which aspects should be guiding in making these end-of-life decisions?

An essential difference between human medicine and veterinary medicine is the issue of informed consent. Where in human medicine a patient must consent to a specific therapy, animals just cannot. I will discuss this topic and the role of both owner and veterinarian in the decision-making process. In my opinion, the decisive factor should be the animal's (future) quality of life. In this context, I define quality of life as the wellbeing of the animal in relation to the gained lifespan relative to the age of the animal but also to the amount and the duration of suffering associated with disease and treatment. I will show that a deliberate estimation of the future quality of that life should be an essential step in making the right choice. Next, I address the role veterinarians play in guiding pet owners to make well-informed decisions about ending the life of their pets.

The first part of this chapter addresses the question of how to decide between continuing veterinary treatment and euthanasia. I analyse the special position of companion animals compared to other (domesticated) animals and reflect on some advantages and disadvantages of the so-called human-companion animal bond. Next I zoom in on consequences for end of life decisions in relation to veterinary medicine. Pet owners as well as veterinarians, each have their own motives and beliefs to extend the life of animals. Unfortunately, in cases of 'heroic medicine' the interests of the animal in question are not always on top of the list.

The second part focuses on the moral implications of our relation with companion animals regarding the practice of convenience euthanasia. Although many pet owners value their beloved pets as members of the family, this is not always reflected in consequent behaviour. Apart from fundamental objections against certain treatments like chemotherapy or serious lack of financial means, some pet owners are simply not prepared to pay the price for necessary veterinary care. Some will rather abandon their pets when taking care of them becomes problematic. For instance in the case of behavioural problems or when they come to the conclusion pet ownership is not really their thing. For this reason animal shelters are structurally overcrowded and yearly a considerable number of healthy, mostly older pets have to be euthanized. This raises additional ethical issues. Should we euthanize companion animals for the owner's convenience? Can lack of finances or unwillingness to finance certain therapies be morally legitimate for euthanizing companion animals? Is there any justification for killing healthy companion animals?

To address these issues, I explain the concept of euthanasia in veterinary medicine. Then I will give some arguments that are used for social acceptance of euthanasia of companion animals. If the interests of our companion animals are guiding and their quality of life is central, only in cases of unbearable and hopeless suffering it is easy to morally accept euthanasia. Unfortunately there are many other reasons for people to request for euthanasia. I will show that veterinary guidelines for euthanasia can be helpful, but are not a panacea. Finally, I aim to find arguments for the justification of euthanasia of companion animals even when they are considered healthy.

13.2 The human-companion animal bond

If you consider the status of domestic animals in society, companion animals take a special position. At least in modern Western societies almost every pet owner will happily declare that his pet is regarded as a family member, with corresponding moral standing (Endenburg, 1991). You would expect the interests of companion animals to be adequately covered then.

Humans have a special relationship with their pets, which is often referred to as the human-companion animal bond (Knight and Herzog, 2009). This bond is built on friendship, loyalty and love. Pets seem to give humans unconditioned companionship. There even is evidence that companion animals improve the mental and physical health of their owners (Arhant-Sudhir *et al.*, 2011). And although humans have kept animals as companions for ages, in our present industrialized and individualistic society pets often function as surrogates for human companionship (Shepard, 2008).

On the other hand the human-companion animal bond also has some features of reciprocity. In return for their companionship pet owners provide their animals with food, shelter, attention, veterinary care and so on. Because of these aspects of reciprocity it almost seems fair to speak of a social contract between humans and their companion animals. An unequal contract however, in which humans act as moral agents and animals as moral patients. Logically when the contract is breached it is always from the human side. Unfortunately there are plenty examples of this. We breed companion animals according to our beauty standards, like brachycephalic dogs, with serious consequences for their health. We retain them basic needs like sufficient exercise and feed them in a way that they become obese. The human-companion animal bond is clearly not an equivalent relation because animals are in many aspects depending on their owners. The starting point of the bond is that we as humans decide to keep companion animals. Humans determine to maintain or break the bond and the animal's life is largely controlled by preferences of its owner. It is therefore just to state that, although we keep them in our homes and share our life-styles with them, to keep animals for companion is actually just another form of animal use (Sandoe and Christiansen, 2008).

There is little specific welfare regulation on companion animals in the Netherlands as well as in most other European countries. Companion animals are situated in the

Informed consent is a cornerstone of biomedical ethics. It is founded in the ethical principle of patient's autonomy (Beauchamp and Childress, 1979). The purpose is to prevent patients from being treated against their will. Before applying burdensome therapies like chemotherapy on patients' informed consent, is crucial. What about informed consent in veterinary medicine? Because animals are not able to, in veterinary medicine it is normal to ask the owner to consent to a treatment. Historically, when the veterinary profession was primarily serving agriculture, informed consent was based on preserving the animal's economic value for the owner. In companion animal medicine this pragmatic approach based on a cost-benefit analysis changed to a more moral one, comparable with the process of informed consent in paediatrics. The objective of informed consent in companion animal medicine is to provide adequate information so pet owners can make a responsible decision for their pet and for themselves. This also applies for end of life decisions. In case of severe illness it's eventually the owner who decides what happens.

In this process veterinarians have to take into account the interest of both animal and owner and the bond they have together. The risks and benefits of all treatment-options, including palliative care and euthanasia, need to be considered. Trust and good communication skills are essential conditions to come to a deliberate decision. In veterinary oncology, for instance, it must be clear what the desired outcome of the treatment is. Since cure is not always possible, expectations of owner and veterinarian about successful remission and quality of life after treatment must be clear. Side effects and recurrence of cancer should not be a surprise for the owner. The veterinarian doesn't have to maintain absolutely neutral in his advice. The focus must be on the health and welfare of the animal in question. In this respect he should also take into account the possible side effects, corresponding suffering and the duration of the treatment in relation to the expected gain of lifespan. He should do this without imposing personal preferences resulting in informed consent against the owner's own inclinations (Fettman and Rollin, 2002).

13.4 Quality of life

A number of aspects have contributed to the fast development of highly sophisticated veterinary medicine over the last decades. The evolution of technical possibilities derived from human medicine, the special moral status of companion animals, the prosperity in Western societies and the increased societal sensitivity for animal suffering are the most important ones. As a result of this it has become possible to prolong the life of our pets with many years. In human medicine, besides prolongation of life, the emphasis is more and more on the future quality of life. A longer life is not necessary a better life. As humans we are (to some extent) able to consider the consequences of treatment for the future quality of our life. As long as we are well informed and mentally and physically competent, we can balance risks and benefits and autonomously decide about treatment options.

Because of evolutionary, behavioural and neurological analogues between humans and animals it is plausible that many animals have conscious experience of pain and suffering. Some animals are even capable of understanding basic causality. According to Bermudez though, animals don not have the capacities for second order thoughts (thinking about thinking) because they lack the necessary linguistic ability (Bermudez, 2007). Thanks to higher order thinking humans endure pain and suffering on short notice because they value long-term goals. It is highly questionable if animals have the cognitive possibilities to weigh future benefits against current misery. One could even say that animals suffer more than humans in a comparable condition of illness. Animals probably don't have the capacity to understand why they suffer, which can provide considerable relief in the case of humans (cf. Nuffield Council on Bioethics, 2005). Humans are also able to distinguish physical suffering from mental suffering and their experienced quality of life. A severely paralysed person can consider his life very worthwhile living. It is believed that animals cannot weigh these different kinds of suffering. A paralyzed dog is therefore likely to suffer because it is deprived of the possibilities to exhibit normal canine behaviour. It is deprived of the possibility of being a dog. This implies the owner as well as the treating veterinarian has a responsibility to safeguard the animal's quality of life during treatment, especially when this treatment is long and stressful. Moreover, they have to make a deliberate decision whether this treatment can be justified in first instance, taking into account the expected future quality of life of the animal.

Because the lack of higher order thinking it's unlikely that animals can fully understand the concept of life and death. Presently experienced quality of life is probably their major concern. In this sense animals probably also don't know the concept of hope for a better future. You could argue though, that owner and veterinarian can more or less predict an animal's future quality of life after treatment. So it would be strange to deprive an animal of future happiness, just because an animal itself cannot oversee the consequences of treatment. In fact the situation is comparable with human medicine where parents and doctors decide about treatment of children.

If quality of life is the most important factor in decisions about veterinary treatment, how can we define what quality of life for a specific companion animal entails? From Nagel we have learned that we shall never know 'how it is like to be a bat' (Nagel, 1974). It is therefore simply impossible to grasp the quality of life of an animal objectively. Although in humans this is a subjective concept as well, the fact that animals are not able to express their opinions about the quality of their life makes it even more complex. Inevitably, an animal's quality of life is an anthropogenic interpretation of both owner and veterinarian, but it is the best we have to offer. Together with the veterinarian's expertise, the owner's impression of the wellbeing of animal is crucial to find out what an acceptable quality of life is. Veterinarians as independent professionals are trained in signalling and assessing signs of animal discomfort. Animal owners maybe have a more subjective impression but generally know their animals best and can determine whether an animal is behaving abnormal. Both findings contribute in establishing actual quality of life of the animal.

But how do we define an animal's quality of life, then? In my opinion the concept of quality of life of an animal has a strong connection with the physical and mental wellbeing of an animal including the possibility to fulfil its species-specific needs. Terms like quality of life, wellbeing or welfare are often used interchangeably. I believe that when we speak of quality of life, we often mean that an animal has to be in good welfare. Therefore, to understand what an animal's quality of life entails, it is useful to consider the concept of animal welfare first.

According to Fraser three biological views on animal welfare can be distinguished (Fraser, 1997). The first perspective on animal welfare is that an animal is free from diseases and able to function well, for instance to grow and (re)produce. Farmers, who believe that if they take good care of their animals, they grow and produce well and therefore are in a state of good animal welfare, often hold this view (De Greef *et al.*, 2006). A second perspective is that animals have a good welfare if they are feeling well. In this view animals are regarded as sentient beings with the possibility to experience pain and pleasure. These two views can be related with the utilitarian approach of maximizing preference satisfaction. When an animal can satisfy its preferences, regardless of what these preferences exactly are, an animal will have positive feelings and is in a state of good welfare. A comparable relation with utilitarianism could also apply for the first perspective, where from the viewpoint of the farmer preferences are satisfied. The third perspective on animal welfare emphasizes the development of species-specific characteristics. In this view, an animal is in good welfare only if it can fulfil its species-specific needs and can perform its natural behaviour. This approach can be connected with, for example, the Aristotelian based capabilities theory of Nussbaum (2004). In practice these views often overlap. I therefore believe that a good estimation of the quality of life of an animal should contain all three aspects. A good quality of life means that an animal must function well, feel well and must be able to fulfil its species-specific needs.

So quality of life must not be restricted to absence of physical pain. When after treatment the animal is unable to perform natural behaviour or species-specific characteristics like running, playing, interacting with other animals and humans, this could lead to serious mental stress and would affect its wellbeing negatively. In these cases future quality of life can be considered low and veterinarian and owner should reflect on the question if such a life is a life worth living for the animal or whether euthanasia might not be a better option.

When quality of life is assessed the duration of the treatment and the expected lifespan animal are relevant as well. A very intensive and long lasting therapy might be considered in a young animal but should be applied with reticence in animals of old age. In relation to end of life decisions in veterinary medicine I will therefore interpret the concept of quality of life as the wellbeing of the animal in relation to the gained lifespan relative to the age of the animal.

Sometimes an owner might think that suffering because of treatment is an acceptable price for some extra time with his animal. But driven by anthropomorphism the owner may have a wrong conception of animal welfare. This can lead to a paradox where people who sincerely care for their animals can go too far and lose sight of the animal's quality of life, while trying to prolong their time together. In these situations the veterinarian must operate as the animal's advocate and guard against keeping a suffering animal alive too long (Rollin, 2007a).

13.5 Role of the veterinarian

According to Rollin the most fundamental question in veterinary ethics is whether the veterinarian has prior allegiance to the client or to the animal. Illustrating this he compares two extreme business models for veterinarians: the car mechanic or the paediatrician. According to Rollin, most veterinarians by nature incline to the latter model (Rollin, 2007b). But running a veterinary practice is also running a business. A veterinarian is financially depending on the services he can provide to his clients. This could be a disturbing factor in the decision-making process and may influence the veterinarian's role as animal's advocate. Prolonging an animal's life with extensive treatments can be financially prosperous. Beside this, the public nowadays critically watches all so-called 'notable professions' and there is a disciplinary court in place to correct disgraceful conduct. However, in cases of exaggerated 'heroic medicine' the owner might not always complain and because animals simply cannot, the profession has a social and moral duty to regulate itself and address the issue of possible overtreatment. In this context McKeegan suggests that a veterinarian's ethical priorities should be (in the following decreasing order): maintain and improve the quality of the animal's life, to extend the quantity of life in relation to the animal's age, to serve owners, to help to contribute to a profitable business or build on a vet's own business and to develop new treatments (D. McKeegan, personal communication).

This leads to the question how the veterinarian ought to fulfil his task as an animal's advocate. A good start is trying to achieve agreement with the pet owner over the fact that the fundamental basis of veterinary treatment is quality of life of the animal. The outcome of good veterinary care should be that an animal is functioning well, feeling well and is able to perform its natural behaviour. Obviously, there will be gradations. In a dialogue the veterinarian and the owner should establish what the threshold of quality of life for a specific animal should be. Ideally, this consensus should be reached before an owner becomes a client of the veterinary practice. These kinds of questions can be part of an intake procedure for new clients. On that basis a veterinarian can educate the owner about the animal's welfare.

Whenever serious illness requires intensive therapy, the purpose must be to start a dialogue wherein quality of life and suffering are continuously weighed. When quality of life is seriously compromised the veterinarian should discuss the option of euthanasia. In this interaction the veterinarian should make use of his Aesculapean authority, defined by Rollin as the unique authority that is vested in every healer, without pushing the owner in a certain direction too much (Rollin, 2002).

Also helpful in making end of life decisions are the basic principles of biomedical ethics that Beauchamp and Childress introduced: beneficence, non-maleficence, justice and respect for autonomy (Beauchamp and Childress, 1979). Autonomy means enabling patients to make a rational informed choice and to respect this decision. It was integrated in human medicine to secure voluntary informed consent, protect confidentiality and avoid discrimination and abuse in medical practice. Respect for autonomy is difficult to implement in animal ethics because animals lack the cognitive and communicative capacities to give informed consent. However, when interpreted as the possibility and ability to perform species-specific behaviour. I think the principles of biomedical ethics, including autonomy, can be useful in the context of euthanasia of companion animals. Mepham, for example, has also used these principles in his ethical matrix to address ethical issues in biotechnology (Mepham, 2000). His matrix is a tool to identify the different interests of stakeholders involved in ethical issues, including animals. Mepham distinguished three main principles: respect for well-being, autonomy (to perform species specific behaviour) and justice (or fairness). Mepham's tool can be helpful for veterinarians in making decisions about killing companion animals as well.

13.6 The moral limits to veterinary medicine

The purpose of the first part of this chapter was to find an answer to the question: where to draw line in companion animal veterinary medicine? When do we decide for treatment and when is it morally acceptable to end the life of companion animals? After going through the different ethical issues involved, I come to the conclusion that there certainly are limits, but it's impossible to draw a clear line. The boundaries in veterinary medicine are not so much set by the kind of treatment but by the interests of the animal in question. This change of focus can also be applied to end of life decisions in companion animal medicine. Because we give our companion animals a certain moral standing and consider them members of our family, we must take their interests very serious when making decisions about veterinary treatment or ending their life. In this aspect quality of life is more important than quantity of life or prolonging the human – companion animal bond. Therefore the interests of the concerned animal should outweigh the interests of the owner.

It's of course very difficult to give an exact account of what an acceptable quality of life is for a specific animal. The amount of physical pain is surely not the only factor. The ability to perform as much natural behaviour as possible is equally important. When, after treatment, a companion animal is no longer able to function as a normal member of its species, this can cause serious mental suffering as well. Therefore I think it is problematic to consider such a treatment as morally justified. Because animals aren't able to relate the possible future benefits of a long and intensive treatment to current suffering, veterinarians should be careful in advising highly advanced veterinary care without the perspective of short recovery time, sufficient possibilities of nursing and palliative care and eventually a good prognosis. Euthanasia can be a better option then.

There may also be circumstances in which a physical disorder that doesn't affect an animal's welfare directly but that does make the animal unfit for its purpose. An example of such a situation is a dog with permanent urinary incontinence. Although such a dog may not experience this disorder as a serious compromise of its quality of life, one can imagine that such an animal is difficult to keep indoors. This corresponds with an animal welfare perspective in which an animal has to function well. When no realistic alternative can be offered in this kind of situations, ultimately euthanasia could be considered.

The best possible way to estimate what an acceptable quality of life entails is to combine the estimation of the owner, who knows the animal best, with the expertise of the veterinarian. The veterinarian has to act as the animal's advocate and must provide the owner with all the necessary information to come the right decision. This is how informed consent in veterinary medicine must function. Veterinarians have to educate pet owners about the differences between humans and animals to prevent anthropomorphism that could harm the interests of animals. They have to be trained in making these difficult decisions together with the owner. Besides technical knowledge and knowledge about species-specific animal welfare, trust and communication-skills are essential to come to a deliberate decision.

13.7 Assessment model for companion animal euthanasia

The word euthanasia is derived from the Greek and means literally the good ('eu') death ('thanatos'). Euthanasia normally refers to the medical practice of intentionally ending a human life in order to relief pain and suffering. In the context of animals, the concept of euthanasia is used commonly when animals are killed painlessly, regardless of the motivation. This implies a veterinarian is requested to actively kill an animal.

With regard to euthanasia of companion animals two considerations are notable. The first question is whether killing is in the interest of the animal itself and his future wellbeing. The second consideration is whether there are human interests that outweigh the interest of the concerned animal to stay alive. This last question gives of course most ground for debate. There are in fact two kinds of human interests. One can have an individual interest, like the financial interest of a pet owner who has to pay the bill for veterinary care. These interests can be controversial. Normally society considers the killing of an animal with a treatable disease unacceptable, unless there are good (psychosocial) reasons. For instance, half of the Dutch population and a large majority of vets consider euthanasia of companion animals unacceptable, when the owner is simply not prepared to pay for the veterinary costs (Rutgers *et al.*, 2003). On the other hand there are broader societal interests, for instance economic, political or public health interests in case of a zoonotic disease outbreak. A serious risk for general public health is generally regarded a valid reason for killing animals. Human interests have to be considerable high, though, to justify killing of companion animals in our society. It must be made clear that there are no reasonable alternatives that could prevent the killing (Rutgers *et al.*, 2003).

Balancing these interests can be difficult. When making end of life decisions, veterinarians and owners have to find a morally acceptable justification for euthanasia. Society entrusts the veterinary profession with the delicate task to weigh requests to perform euthanasia and to give the owner advice in end of life decisions. To help individual veterinarians we have developed an assessment model to give them moral guidance. Furthermore, this guideline makes the decision procedure of veterinarians more transparent and shows society how the professional responsibilities of veterinarians are fulfilled.

13.8 Unbearable suffering

Veterinarians often experience moral stress when they are asked to euthanize animals (Rollin, 2011). Because animals cannot decide for themselves to take the lethal drugs and pet owners are not allowed to administer them, it always comes down to veterinarians to kill them. Moral problems most often occur when it concerns healthy animals or animals that can be treated easily and without high costs. On the other hand, veterinarians are confronted with pet owners who refuse euthanasia even if the animal is suffering unbearable and hopelessly, as I described earlier.

Veterinarians can refuse euthanasia when the animal's health and welfare is not seriously compromised. Normally veterinarian and pet owner will decide together whether euthanasia is indicated. Criteria like pain and suffering, treatment options, duration of treatment, prognosis and future quality of life in relation to expected lifespan, can be used to make a deliberate decision. But also cost of veterinary care, the possibility to provide the necessary extra care at home and the future fitness of the animal for the purpose it is used, play a role. The primary focus of the veterinarian in this process should be on the quality of life of the animal. As discussed above, animal welfare should be not only defined as functioning and feeling well. Animal welfare is more than absence of pain and suffering. The (future) possibility for an animal to perform their species-specific behaviour is of equal importance.

When confronted with a request for euthanasia veterinarians have to weigh the interests of the owner and of the animal. Most of the time these interests point in the same direction, but unfortunately not always. To help veterinarians and pet owners to make these difficult decisions, assessment models can be helpful. Professional guidelines help veterinarians to explain pet owners what is good veterinary practice in case of euthanasia. After all, society expects the profession to make prudent and transparent decisions when it comes to ending an animal's life. Although the professional responsibility to help animals in need and to safeguard animal health and welfare as well as public health, is laid down in law and the professional code of conduct, it not always an easy task to make right decisions about the life and death of animals.

The purpose of our assessment model is to help veterinarians, by guiding them systematically through a number of important questions. The presuppositions of our tool are the interests of the animal that should come first. Quality of life in terms of functioning and feeling well and able to perform natural behaviour is the central parameter. On the other hand, public health interests can trump interests of individual companion animals.

In the first step of our assessment model the veterinarian examines if the animal has a physical disease. If so, he has to assess the gravity of the disease and the involved suffering. When suffering is unbearable and hopeless it is the moral and professional duty of the veterinarian to euthanize the animal. Sometimes the owner has to be convinced this is best for the animal. When the animal is not suffering unbearable neither hopeless, the following question is if there is an underlying disease that has to be treated. Some diseases do not need special treatment nor diminish the quality of life. Euthanasia is not an option then.

When a disease is treatable, the veterinarian should present all possible treatment options. As I discussed earlier, animal's interests must be leading. A pet owner, however, also takes things like veterinary costs and possibilities for extra care at home into consideration. This can make them decide for euthanasia, which puts moral pressure on veterinarians who normally will be reluctant to euthanize an animal with a treatable disease. Within reasonable limits pet owners have to understand that responsible pet ownership entails veterinary costs and extra care when needed. It is difficult to define what these limits are in this context. Probably, this can only be established on a case-by-case basis.

When money is an issue, veterinarians are often prepared to make special arrangements for payment. In the Netherlands there are also several public and private funds or animal welfare organisations that can help people to finance veterinary care. To prevent discussions about cost of veterinary care pet insurances can provide a solution. Naturally this is not an option when the animal is already ill. In some cases when the best veterinary treatment is financially infeasible, it is possible to provide a less costly alternative that is still acceptable in terms of animal wellbeing. Under certain conditions, for instance, conservative treatment of an orthopaedic fracture in, for example, cats and rabbits can be a responsible choice. Bottom line is that all options have to be considered carefully before euthanasia comes into the picture. In these situations veterinarians can help to look for alternatives, but ultimately the owner is in the lead to look for a satisfactory solution in the animal's interest. Alternatives for euthanasia could be to renounce the animal to a new owner or an animal shelter. Sometimes it is possible to provide alternative and less expensive veterinary care, but only when the quality of life of the animal is secured. Some diseases, however, cannot be cured and do impair the wellbeing of the animal. In these situations it depends on long-term prognosis and quality of life whether euthanasia is indicated or not.

It frequently occurs that pet owners request for euthanasia because of behavioural problems. A survey amongst veterinary practices in the UK revealed that 5.9% of the euthanized dogs and 1% of the cats were euthanized for behavioural problems (Edney, 1998) Certainly in case of aggressive animals veterinarians have to assess if the animal is a threat for its surroundings, especially in families with children. If so, the next question is whether this behavioural problem is treatable. Veterinarians are expected to make a good risk assessment in the interest of public health. When the risk is too high, it is in the interest of society to euthanize the animal before serious damage is done. Other cases of companion animal behavioural disorders like incontinence, vandalism, noise or separation anxiety are in first instance eligible for treatment and transfer to an animal shelter or a new owner when appropriate.

The interests of the animal are the model's starting point. When an animal is not suffering unbearable then euthanasia cannot be justified in advance. All reasonable alternatives must be examined before a veterinarian can decide to euthanize an animal that is not suffering unbearably. Whether the search for alternatives is successful

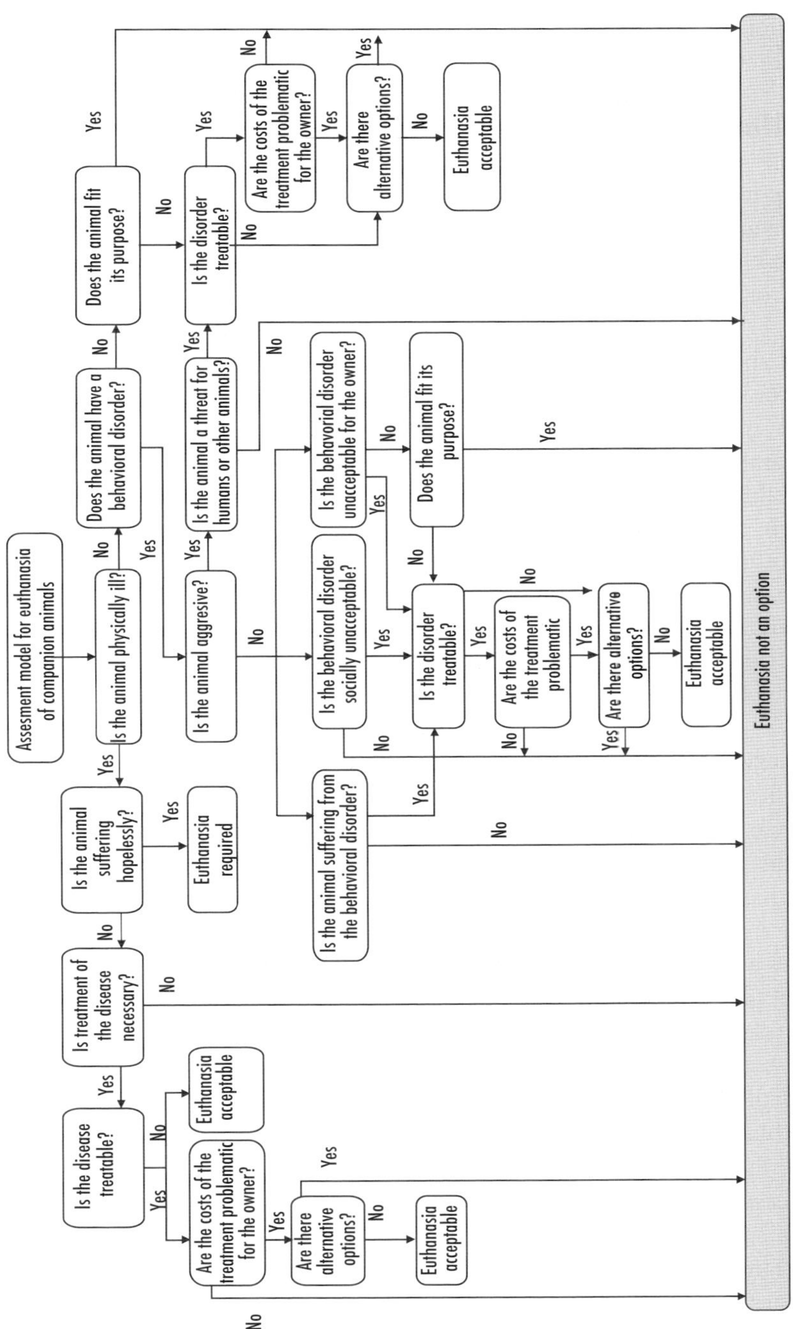

Figure 13.1. Assessment model for euthanasia of companion animals.

depends not only on the veterinarian's advice but also on the effort and willingness of the animal owner to invest in alternative solutions.

13.9 Killing healthy companion animals

According to the general opinion in Europe it is justified to kill companion animals that suffer unbearably and hopelessly. In fact, in many countries one is even legally obliged to kill animals in such cases. In this sense euthanasia of companion animals is hardly controversial.

But what about euthanasia of healthy companion animals? In veterinary practice requests for killing healthy animals is common. In the light of the special position we give them, it seems hard to imagine cases where killing healthy companion animals can be morally justified. Perhaps it is explicable from a utilitarian point of view, when killing healthy animals leads to maximization of the overall greatest happiness. For instance, in situations in which public health is endangered by threat of a zoonotic disease outbreak. But in these rare occasions maybe even an animal rights protagonist like Regan would justify killing healthy companion animals. Although Regan grants animals and humans equal moral rights, in cases of self-defence, he thinks human interests trump those of animals (Regan, 1983). Besides zoonotic diseases this could also be true in case a very aggressive dog, for instance.

In veterinary practice, pet owners sometimes ask for euthanasia because their animal does not fit its purpose anymore. This happens with hunting dogs or breeding dogs, for instance. Though in some cases these animals are not qualified as companion animal, the possibility to give them a good home should be seriously considered before euthanasia is discussed. Other examples in this category are a litter of unwanted kittens or a change in the family situation that makes the pet a burden. In many of these cases a satisfactory solution can be found and euthanasia will be the last resort. These examples show that there is a great need to educate society about responsible pet ownership. When we believe it is acceptable to keep animals as companions, we should also face the corresponding responsibilities. As I mentioned in the first part of this chapter the special duties derived from the human-companion animal bond demand this.

In these cases of convenience euthanasia the interest of the owner does not trump the animal's interest. Euthanasia for reasons of convenience is morally unacceptable. Veterinarians should refuse euthanasia on these grounds. To resist the pressure of an owner's insistent request and to standardize the decision making process of veterinarians we have developed the aforementioned assessment model for euthanasia of companion animals. This model helps individual veterinarians to make deliberate end of life decisions but also improves the accountability of the profession towards society concerning the practice of euthanasia of companion animals. The force of this assessment model is that it supports veterinarians to underpin their decision and gives them the confidence this decision is backed up by their entire profession.

Another example is the killing of healthy animals that takes place in animal shelters. To avoid euthanasia in companion animal practice, transferring animals to an animal shelter seems to be a reasonable alternative. However, animal shelters are often overcrowded with animals for which people, for different reasons, can no longer provide the necessary care. And for some of these animals a new and suitable home is not easy to find. For this reason, a great number unadopted or unadoptable animals are killed yearly. In animal shelters in the Netherlands yearly about 7.6% of the dogs and 9.2% of the cats are euthanized (NVBD, 2014). Several arguments are brought forward to justify this practice. One is to say euthanasia is in the interest of those animals because a lifelong stay in a shelter is considered worse off. Another is humane killing of those animals is necessary because of overpopulation of animals in shelters. Finally, it is argued that killing surplus companion animals in shelters reduces the growth of feral populations of cats and dogs who could potentially endanger public health as a source of infection. However, this last argument may be more relevant in the USA or countries in Eastern Europe but at the moment it is not really an issue in the Netherlands. Basically, these arguments seem to reflect a utilitarian view. Humane killing of a few healthy animals maximizes overall human and animal welfare (Palmer, 2008). Other ethical views disapprove of euthanasia of healthy animals in animal shelters. Regan, for instance, condemns this practice. He thinks an untimely death of animals maybe does not hurt them if this is done painlessly, but these animals will be harmed because all possibilities for future satisfaction are taken away. He doesn't consider this method true euthanasia. Regan thinks that since animals are 'subjects of life' and therefore have the right to be treated equal to humans. To euthanize an animal three requirements have to be fulfilled. The killing has to be painless, it has to be in the animal's own interest and the person who euthanizes the animal must be sincerely motivated to act in the interest of the concerned animal. He calls this preference-respecting or true euthanasia (Regan, 1983).

Besides that, Palmer points out the special dependency-relationship these animals have with us. We have a moral responsibility towards these surplus companion animals, that we have helped creating ourselves. From this responsibility she questions the moral justification of killing healthy companion animals in animal shelters. It is not an appropriate way to discharge our responsibility towards these dependent animals. She suggests we need a greater collective responsibility for the existence of all companion animals. This means a duty of care towards these, often dependent, domesticated animals. According to Palmer, a reason for the current social acceptance of killing healthy animals in animal shelter is our attitude of instrumentalisation towards domesticated animals in general (Palmer, 2008). This endorses Sandoe's view that despite the human-animal bond, to keep animals as companions is actually just another form of animal use (Sandoe and Christiansen, 2008). This conclusion emphasizes the importance of our assessment model for veterinarians to safeguard deliberate decisions about euthanasia of healthy companion animals.

As indicated, veterinarians are often confronted with requests for euthanasia of healthy animals. These requests are mostly difficult to justify. Old age in itself is probably not a good reason for euthanasia. Certainly when an animal seems to function and feel well and is able to perform its natural behaviour it is not in the interest of the animal to end its life. But when this same old animal is doomed to stay in an animal shelter for the rest of its life because no one is interested to give it a new home, you could wonder whether this animal will lead a live worth living. I believe euthanasia of non-rehomable animals in animal shelters could be justified under certain circumstances.

13.10 Don't let euthanasia be the easiest way out

As I have shown, killing of companion animals raises two sorts of moral issues. The first is that there is a risk we sometimes postpone the killing of companion animals causing unnecessary suffering. The second issue is that on the other hand we occasionally kill them too soon, depriving them a natural lifespan and possible future wellbeing.

Although pet owners sometimes go too far in the love for their animals and want to keep them alive at all costs, misuse of euthanasia to solve problems with unwanted companion animals seems a greater problem and happens on a larger scale. From a moral point of view, for most people unbearable and hopeless suffering is an undisputed ground for euthanasia. This suffering is not necessarily physical. Mental problems or the impossibility to perform species-specific behaviour can also impair quality of life seriously. In cases of untreatable behavioural disorders, for instance, euthanasia can be the best option.

Another dilemma is whether there is a socially accepted limit to the amount of money an animal owner is required to spend on veterinary care? When is it justified for an owner to stop the treatment for financial reasons? I have not elaborated that issue. Perhaps it is not the main concern of the veterinarian to worry about these financial aspects of veterinary care. Nor can he determine whether the costs for veterinary care are to be considered reasonable. Ultimately, the decision to continue the treatment is up to the owner. However, veterinarians could point at the moral responsibility of pet owners to provide their animal the necessary veterinary care.

There are no easy answers to the two main moral issues I have presented. Justification for euthanasia is often only possible on a case-by-case basis. Professional assessment models help veterinarians to make a deliberate decision. When there is no mental or physical pain and suffering involved euthanasia must be considered the last resort. All alternatives have to be examined first. The primary responsibility for exploring the possible alternatives lies with the animal owner. To make deliberate end of life decisions for companion animals, pet owners need to be educated by government, animal welfare organizations and veterinarians on responsible pet ownership.

Veterinarians have a crucial role in the decision-making process concerning euthanasia of companion animals. Because we animals are not able to promote their interests, society has entrusted the veterinary profession with the responsibility to safeguard animal welfare. Veterinarians must place the interests of the animal in the centre and focus on the (future) quality of life as the decisive factor in end of life dilemmas. Generally the animal's interests predominate.

When advising on end of life decisions veterinarians have to weigh all interests in play, including their own. On a case-by-case basis veterinarians must give sound advice by making good use of their professional authority. However, there can be individual or societal human interests that could outweigh the interest of the animals in question. A serious threat of public health is the clearest example. Convenience euthanasia on the other hand has to be condemned. Veterinarians have to be critical and actively guide pet owners to look for alternatives. Because of the special status we as society give our companion animals and their dependency on us, that exist only because we have created them, euthanasia cannot be an easy way out.

References

Arhant-Sudhir, K., Arhant-Sudhir R. and Sudhir, K., 2011. Pet ownership and cardiovascular risk reduction: supporting evidence, conflicting data and underlying mechanisms. Clinical and Experimental Pharmacology and Physiology 38: 734-738.

American Veterinary Medical Association (AVMA), 2012. Guidelines for responsible pet ownership. Available at: http://tinyurl.com/ofatk8v.

Beauchamp, T.L. and Childress, J.F., 1979. Principles in biomedical ethics. Oxford University Press, New York, NY, USA, 480 pp.

Bearup, B.J., 2007. Pets: property and the paradigm of protection. Journal of Animal Law 3: 173-191

Bermudez, J.L., 2007. Thinking without words, an overview for animal ethics. Journal of Ethics 11: 319-335.

Cooke, S., 2011. Duties to companion animals. Res Publica 17: 261-274.

De Greef, K., Stafleu, F. and De Lauwere, C., 2006. A simple value-distinction approach aids transparency in farm animal welfare debate. Journal of Agricultural and Environmental Ethics 19: 57-66.

Dutch Civil Code, 2015. Book 3, article 2. Available at: http://tinyurl.com/oropjld.

Edney, A.T.B., 1998. Reasons for the euthanasia of dogs and cats. Veterinary Record 143: 114.

Endenburg, N., 1991. Animals as companions: demographic, motivational and ethical aspects of companion animal ownership. Thesis Publishers, Amsterdam, the Netherlands, 144 pp.

Fettman, M.J. and Rollin B.E., 2002. Modern elements of informed consent for general practioners. Journal of the American Veterinary Medical Association 221: 1386-1393.

Fraser, D., Weary D.M., Pajor E.A. and Milligan B.N., 1997. A scientific conception of animal welfare that reflects ethical concerns. Animal Welfare 6: 187-205.

Knight, S. and Herzog, H., 2009. All creatures great and small: new perspectives on psychology and human-animal interactions. Journal of Social Issues 65: 451-462.

Macdonald, V., 2009. Chemotherapy: managing side effects and safe handling. Canadian Veterinary Journal 50: 665-668.

Mepham, B., 2000. A framework for the ethical analysis of novel foods: the ethical matrix. Journal of Agricultural and Environmental Ethics 12: 165-76

Nagel, T., 1974. How is it like to be a bat. Philosophical Review LXXXIII: 435-450.

National Cancer Institute, 2012. Side effects. Available at: http://tinyurl.com/pdnyz9g.

Nederlandse Vereniging tot Bescherming van Dieren (NVBD), 2014. Jaarverslag 2013 Dierenbescherming. Available at: http://tinyurl.com/njlwehb.

Nuffield Council on Bioethics, 2005. The ethics of research involving animals. Nuffield Council on Bioethics, London, UK, 376 pp.

Nussbaum, M.C., 2004. Beyond compassion and humanity: justice for non-human animals. In: Sunstein, C.R. and Nussbaum, M.C. (eds.) Animal rights: current debates and new directions. Oxford University Press, New York, NY, USA, pp. 299-320.

Palmer, C., 2008. Killing animals in animal shelters. In: Armstrong, S.J. and Botzler, G. (eds.) The animal ethics reader. 2nd edition. Routledge, New York, NY, USA, pp. 570-577.

People's Dispensary for Sick Animals (PSDA), 2011. Animal wellbeing report 2011 – the state of our pet nation. Available at: http://tinyurl.com/pa8gf96.

Regan, T., 1983. The case for animal rights. University of California Press, Berkeley, LA, USA, pp. 113-114.

Rollin, B.E., 2002. The use and abuse of Aesculapian authority in veterinary medicine. JAVMA 220: 1144-1149.

Rollin, B.E., 2007a. Ethical issues in geriatric feline medicine. Journal of Feline Medicine and Surgery 9: 326-334.

Rollin, B.E., 2007b. An introduction to veterinary medical ethics, theory and cases. second edition. Wiley-Blackwell, Hoboken, NJ, USA, pp. 27-28.

Rollin, B.E., 2011. Euthanasia, moral stress and chronic illness in veterinary medicine. Veterinary Clinics of North America Small Animal Practice 41: 651-665.

Rutgers, L.J.E., Swabe, J.M. and Koolmees, P., 2003. Het doden van dieren: maatschappelijke en ethische aspecten [killing animals, societal and ethical aspects]. Wageningen Academic Publishers, Wageningen, the Netherlands.

Sandoe, P. and Christiansen, S.B., 2008. Ethics of animal use. First edition, chapter 8. Wiley-Blackwell, Chichester, UK, 187 pp.

Shepard P., 2008. The pet world. The animal ethics reader. Second edition. Routledge, Oxon, UK, pp. 551-553.

Warren M.A., 1997. Moral status: obligations to persons and other living beings. Oxford University Press, Oxford, UK, pp. 234-235.

Part IV.

Between wild and kept

14. Beneath the surface: killing of fish as a moral problem

B. Bovenkerk[1*] and V.A. Braithwaite[2,3]

[1]Wageningen University, Philosophy, P.O. Box 8130, 6700 EW Wageningen, the Netherlands; [2]Institute for Advanced Study Berlin, Wallotstraße 19, 14193 Berlin, Germany; [3]Center for Brain, Behavior and Cognition, Pennsylvania State University, University Park, PA 16802, USA; bernice.bovenkerk@wur.nl

Abstract

Are we morally justified in killing fish and if so, for what purposes? We do not focus on the suffering that is done during the killing, but on the question whether death itself is harmful for fish. We need to distinguish two questions; first, can death be considered a harm for fish? And second, if it is a harm, how much of a harm is it? In order to answer the first question, we explore four lines of reasoning: (1) fish desire to stay alive; (2) something valuable is lost when fish are killed; (3) death deprives fish of future happiness or goods; (4) killing fish reflects badly on our character. Some argue that we should not kill animals if they desire to stay alive and that a being can form a desire to stay alive only when it has the capacity to be aware of itself as a distinct entity existing over time. We cast doubt on this view: Do we value continued life because it is desirable or do we desire continued life because it is valuable? It seems more plausible that it is not the desire to live that matters, but being able to enjoy goods, and death thwarts future opportunities for enjoyment. This would entail that a being can have an interest in continued life, without actively being interested in it. Next, we discuss the second question of *how* harmful death is for fish. A widely shared intuition is that it is worse to kill a human being or mammal than a fish, because human or mammal life is in our view more valuable. But how can we account for this intuition? Finally, we address some implications of the view that killing fish is harmful.

Keywords: fish, harm of death, desire-account, foregone opportunities account

14.1 'Fish are friends, not food'

In the popular children's film 'Finding Nemo' a group of vegetarian sharks live by the motto 'Fish are friends, not food'. They even attend meetings reminiscent of Alcoholics Anonymous to help them give up fish. While this film demonstrates the absurdity of forbidding a shark to eat fish, it does raise the question of whether *we* are allowed to do so. Should we consider fish our friends and if so, does this mean we should not kill them for consumption, experimentation or recreational purposes? The central question of this chapter is whether we are morally justified to kill fish and if so, for what purposes. This question is distinct from the question whether we are allowed to

*Franck L.B. Meijboom and Elsbeth N. Stassen (ed.) **The end of animal life: a start for ethical debate***

DOI 10.3920/978-90-8686-808-7_14, © Wageningen Academic Publishers 2016

make fish suffer when they are killed. We are, in other words, not primarily focussing on the harm that is done during the killing; rather our question is whether death itself is harmful. We are interested in the question what the value of fish life is, for killing animals is harmful only if animal life is somehow valuable. Of course, we could also argue that killing fish were problematic if it would cause sorrow to or fear in other fish. This could for example be the case when there is a strong bond between two fish, such as mates or mother and offspring. However, for the purposes of this chapter we leave questions of indirect harm aside.

We will assume that fish can experience pain and distress. This has been argued for a number of species that are commonly used by humans (Braithwaite, 2010; Braithwaite and Huntingford, 2004; Sneddon *et al.*, 2003) and moreover, there is a growing degree of consensus – albeit a recent consensus – about this among marine biologists (Huntingford *et al.*, 2007). In fact, due to the recent knowledge obtained about suffering in fish, efforts are now undertaken to guarantee the welfare of farmed fish and to provide welfare-friendly slaughter methods (Van De Vis *et al.*, 2003; Lambooij *et al.*, 2006). However, all the efforts undertaken to improve the welfare of farmed fish would not be necessary (at least theoretically) if we were to come to the conclusion that killing fish for food were not justified in the first place. And *vice versa*, even if we could farm and slaughter fish without any welfare problems, it may still be wrong to do so, if death in itself is a harm for fish. Moreover, discussions about 'welfare-friendly' fishing gear, or reeling in trawl nets at slower speeds to prevent barometric trauma take on a different meaning if killing fish in itself would be morally problematic.

We need to distinguish two questions: (1) can death be considered a harm for fish? (2) If so, how much of a harm is it? In other words, would it be sufficiently harmful to plea for an end to fish consumption, sports fishing, or other activities where fish are routinely killed, such as in the aquarium industry or animal experimentation? A widely shared intuition is that it is worse to kill a human being or mammal than a fish, because human or mammal life is in our view more valuable. But how can we account for this intuition? We should mention here that when we speak about 'the harm of death', we are talking about harm in a moral sense. Of course death harms a fish in the sense that its body is damaged – in the same sense as a plant can be harmed when it is cut – but is this a harm that matters morally? In the first part of this chapter we address the question whether death should be considered a moral harm for fish at all. As far as we can tell, four lines of reasoning can be put forward to argue that death is a harm for animals in general. We discuss these in the next four sections and examine whether they apply to fish. In the second part of this chapter, we tackle the question of *how* harmful it is, and finally, we briefly describe some practical implications of considering that death is harmful for fish. In particular, we discuss the catch and release versus catch and kill debate, the problem of bycatch in commercial fishing, and the question whether we should stop predators from killing fish. Regardless of our answer to the question of whether we are morally justified to kill fish, it is realistic to hold that fish will be killed in large numbers for the foreseeable future. Therefore, our discussion is by no means meant to replace further debate regarding the question of how to guarantee better fish welfare.

14.2 Fish have a preference or desire for staying alive

When we consider why it would be wrong to kill a person, the first reason that comes to mind is that this person does not want to be killed. The person has a desire or preference for staying alive. Is this a convincing reason and if so, does it apply to fish as well? Whether this is a convincing reason, amongst other things, depends on what moral theory one adheres to. From a classical utilitarian point of view, where all that matters is the total amount of happiness or pleasure in the world, a desire for staying alive is irrelevant. After all, when one is dead, one's desires cease to exist and one can't be harmed anymore; if only experiences count, death doesn't harm one, for when one is dead one ceases to have experiences. As Singer (1980) points out, killing as such is not wrong in classical utilitarianism; it could only be wrong indirectly, when those who stay behind worry about being killed next and this worrying decreases the total amount of happiness in the world. This, however, entails that if someone were killed in secret, death would not be a harm to this person. To avoid this counterintuitive implication, Singer adopts preference utilitarianism, arguing that we should weigh preferences of all beings with interests against each other. When a being has a preference for staying alive, this is an important factor to take into consideration – as it is a very strong preference – even if this preference can ultimately be overruled by other preferences in a utilitarian calculus.

The question for a preference utilitarian is, then: What beings have a preference to stay alive? When confronted with avoidance behaviour of animals that are in danger, such as the struggling for survival of a fish on a hook, at first sight we might interpret this as a fear of death.[1] Singer (1980), however, warns us against taking this to mean a preference for continued existence. Rather, we should interpret this as a desire to stop the pain or the threatening situation and of course this desire can also come about by killing the animal. In his view, a being can form a preference to stay alive only when it has the capacity to be aware of itself as a distinct entity existing over time (Singer, 1980). In other words, one needs to have awareness, albeit in a limited form.[2]

---

[1] And indeed, some philosophers do interpret it as such (Francione, 2006: 240).

[2] DeGrazia (2009) distinguishes between different forms of awareness and self-awareness, without necessarily implying self-consciousness (cf. Droege and Braithwaite, 2014).

It appears that capacities such as self-reflexivity, higher-order thoughts or a Theory of Mind are not necessary for this type of self-awareness.[3]

Other philosophers approach this question from the other side and argue that death is only a harm to those beings that – rather than staying alive – have a preference not to die, or even stronger, not to be killed (e.g. Bracke, 1990; Cigman, 1981). For them, this entails that one must have a concept of death. Although the motivation that all organisms have to survive and not be killed could be considered to be a preference at some level, these philosophers mean a consciously held preference here. DeGrazia (1996: 235) interprets this similarly as Singer interprets having a preference for continued existence, namely by saying that 'having a concept of death involves having a sense of oneself existing over time'. He suggests that in a certain sense all animals that can experience fear have a sense of themselves existing over time, as they have a sense of themselves as the subject of a harm in the future. We can wonder whether this stretches the phenomenon of fear too far, however. Fear is a very ancient affective state that selection seems to have promoted because animals that experience fear (and this may just be the physiological processes without necessarily an awareness of the fear) are more likely to survive dangerous challenges. Moreover, proponents of the 'concept of death' argument likely have a more demanding reading of what it takes to have a concept of death. For example, some assume that having such a concept requires language or second-order beliefs or intentions (Davidson, 1982; Bracke, 1990).[4]

If we look at the 'harm of death question' from a rights perspective, we see that the whole range of positions in this debate can be taken, but that all depend on the interpretation of the criterion of self-consciousness. Regan (2004) argues that an animal has a right not to be harmed – and he regards killing as such a harm – when an animal is a subject of its own life. This requires consciousness, but only limited self-consciousness. Tooley (1972) on the other hand, argues that a being can only have a right to life if it has a desire to live and that only beings who have an awareness of their desire actually have a desire to live. This is because you need to conceive of

---

[3] According to Singer at least some farm animals do not fit this description and therefore do not have a preference for continued life; fish in his opinion do not fit the description. He has argued that in principle these animals could be killed painlessly and be replaced with similar animals with the same amount of pleasure. In a utilitarian calculus the total amount of happiness would not diminish in this case. This so-called 'replaceability thesis' has been criticized for generating counter-intuitive implications and is open to the charge that we could also painlessly kill and replace infants or mentally handicapped people that do not have a preference for continued living either (Pluhar, 1995). This criticism could be deflected when one does not hold a 'total view', but a 'prior existence view' utilitarianism (Višak, 2011). Animals that do fit this description according to Singer (2011) are great apes, probably whales and possibly certain other mammals such as cats and dogs.

[4] Language in this context tends to be perceived as human language. One could argue that many animals communicate, but this communication does not fulfill the criteria of having a language involving symbols, grammar and syntax. Davidson (1982) however, argues that having concepts at all requires being able to make a subject/object distinction, that 'an intersubjective world' is necessary for having a subject/object distinction and this intersubjective world is only possible when one has human language. Thanks to Henkjan Hoekjen for pointing this out.

The end of animal life

yourself as an entity that exists over time to recognise what aspects of life you desire. Unlike Singer, Tooley thinks this requires consciousness of oneself.[5]

Ruth Cigman (1981) also argues that self-consciousness is required when she argues that death is only a harm for beings with the capacity for categorical desires. Her argument goes as follows: It is not necessary for a creature to *actually* have a desire not to die in order to have a right to live, but the creature must at least have the *capacity* for the desire not to die. This rules out the implication that someone who wants to commit suicide would not have a right to life and could be killed at will. When does one have the capacity to have a desire not to die? Cigman draws on Williams' concept of a categorical desire to make the distinction between having the desire to go on living, or simply having desires for which life is instrumental. Life as a categorical desire answers the question whether or not 'one wants to remain alive' (Cigman, 1981: 58). This refers to reasons one may have to go on living, or those things that make a life worth living or give it meaning. Examples are raising children or writing a book.

Considering the foregoing arguments, we should ask what we know about fish capacities for: (1) awareness of themselves as distinct entities existing over time; (2) self-consciousness or second order beliefs and intentions; and (3) categorical desires. As stated before, we assume that fish can feel pain and distress. To be able to experience these, some form of consciousness is necessary. However, this does not yet tell us whether fish have awareness of themselves over time, let alone whether they are self-conscious or have second-order intentions, or categorical desires. In fact, this is not a question that can be answered yet. Not enough research on fish capacities has been carried out. Such research is made difficult by the fact that over 30,000 different species of fish exist, all with their own evolutionary adaptations. Also, it is difficult to design the tests that could tell us what goes on in fish brains. Tests have been carried out in a number of different species that show they have a memory span much larger than the proverbial three seconds (Nilsson *et al.*, 2008). A range of experiments have shown that many species of fish have the capacity to generate complex representations of their environment rather like a mental map (Braithwaite and De Perera, 2006; Ebbesson and Braithwaite, 2012), and there is evidence that certain fish of different species will cooperate to reach a common goal (Bshary *et al.*, 2006). However, this research does not tell us much about awareness or categorical desires. We think it is safe to assume that fish do not have a concept of death, but then again, we wonder how one could ever find out whether fish, or any other animals for that matter, have such a concept.

---

[5] There are complex and ongoing debates among philosophers of mind about the meaning of consciousness and self-consciousness in animals (cf. e.g. Lurz, 2009). Our analysis here is made difficult by the often ambiguous use of these concepts by animal ethicists.

Now what are the consequences of this lack of information? Should we simply conclude that it is highly unlikely that fish are harmed by death or should we, on the contrary, argue that in the absence of sufficient information we need to apply a precautionary principle and act as if death is a harm for fish? We want to take a different route. Even though it seems to be a stretch of the imagination to conceive of self-consciousness fish, one can raise the question whether this capacity is really relevant. Is it necessary to be able to be harmed by death? One can raise doubts about the argument that death is a harm in as far as we desire continued life. Do we value continued life because it is desirable or do we desire continued life because it is valuable? If we value life and therefore desire it, then perhaps the desire itself is not the decisive factor, but rather the value that we place on life. Life can be valuable for a being, regardless of the question whether this being desires it. For example, we think we should not kill a baby, because the baby's life holds value; the baby itself does not self-consciously desire to live, but that seems irrelevant. One could say that we should not kill the baby because at some point in the future it will desire its own existence, but that seems wrong. What if we know the baby will not live past one year old (and at the moment is not suffering)? We do not generally hold that we are justified in killing it, since it has no concept of its own death or has no desire to live or preference not to be killed. In our view, this is no different for animals. Some philosophers argue that it is not the desire to live that matters, but any central desires a being can have that makes killing that being problematic. In other words, 'death thwarts central desires that they do have' (DeGrazia, 2002: 60). According to DeGrazia, this does not require that beings have a desire to stay alive, but rather that they have 'future oriented projects'. However, this adjusted version of the 'desire account' runs into the same problem as the 'desire for continued life account'. Human babies also do not form future oriented projects yet, but we do not generally agree that death is no harm for them and we are justified in killing them.

14.3 Something valuable is lost when fish are killed

We concluded the previous section by arguing that we desire life because it is valuable rather than the other way around and, thus, the 'desire-account' is not correct in our view. Next, we need to ask what is valuable about continued life. This question could be answered either in an objective or in a subjective sense. According to an objective value theory, a certain good is in a being's interest even if this being does not value this good. This would imply that these non-moral goods are valuable in themselves even if nobody experiences them as valuable (Kaldewaij, 2006). Even if fish would not value anything as good in their lives, their death would still be harmful; perhaps death would not be harmful *to* these fish, but their death would still be a bad state of affairs. What criterion would we use to decide whether this state of affairs is bad? Examples of possible criteria could be beauty, or biodiversity. So, in the case of beauty it would be bad to kill fish if this would diminish the beauty in the world. However, if we were to create extra fish and then kill them, this would not diminish the beauty in the world compared to the prior situation. Would this make it all right to kill these fish? This seems counterintuitive. Also, who is to say what fish are beautiful? On colourful tropical fish we might agree, but what about larger, more bulbous species such as the catfish? Similar objections could be raised to other proposed candidates of an objective list of the good.

14.4 Death deprives fish of future happiness or goods

Let's therefore turn to a subjective value theory. From the point of view of the animal in question we could say that the animal derives pleasure from certain goods in its life and this makes that the animal has an interest in the continuation of these goods. According to DeGrazia (2002: 61), 'death forecloses the valuable opportunities that continued life would afford'. This so-called 'foregone opportunities account' of the harm of death follows Nagel (1991) in that it takes life as instrumentally valuable for all beings that can have experiential wellbeing. According to Kaldewaij (2006: 61) a benefit of Nagel's view is 'that it can explain the magnitude of the harm of death: death takes away the possibility of ever experiencing, doing or accomplishing anything you value again'. One could object that animals are not aware of these foregone opportunities. However, this view on the harm of death does not require that individuals are aware of their lost opportunities. A being, it is argued, can have an interest in continued life, without actively being interested in it. As Višak (2015) explains, according to this foregone opportunities account:

> The harm of death is not determined by how much a being *wants* the future he would have had, but by the loss of *value* that the future would have provided for the being. Hence, according to that view, animals can be significantly harmed by death even if they do not desire to go on living, and even if they have no or few future-oriented desires (italics in original).

All a creature needs in order to be harmed by death in such an account is the ability to have experiences that matter to it and that it would be deprived of when dead. Life is then instrumentally valuable for animals to the extent that they can have valuable experiences that make their lives worth living. According to some, this means sentience is a sufficient capacity one must possess to be able to be harmed by death (DeGrazia, 2002; Simmons, 2011), but others pose stricter requirements, such as memory and a sense of the future. Regan, for example, argues that beings need to be subjects of their lives before their experiences and the possible thwarting of these experiences matter to them (Regan, 2004).[6] As animal welfare scientists have shown, animals do not just have simple desires such as eating when they are hungry, but they actually derive pleasure from eating as well and this makes their life worth living (Duncan, 2006).

While the foregone opportunities account seems rather plausible, it does raise a troubling question, namely whether we can really be deprived of something if we do not exist anymore. After all, when we are dead, we don't know what we are missing.[7] This problem has spurned a philosophical debate too complex to discuss at length within the scope of this paper, but that still casts doubt on our preliminary conclusion that death is a harm for fish. In summary, the debate centres on the question whether you can be harmed by something even if you do not experience this harm. According to Nagel (1991) this is possible. He uses the example of betrayal: if you are betrayed behind your back you are still harmed, even though you don't realise you are betrayed. It has been objected that in this case you still could find out that you have been betrayed and suffer harm as a result of that, whereas when you are dead it is impossible to ever experience the betrayal (Silverstein, 1980). Yet, Nagel's claim is that one has unpleasant experiences when betrayed because betrayal is bad and not that betrayal is bad because it generates unpleasant experiences, and therefore a betrayed person that will remain ignorant of the betrayal is still harmed (Nagel, 1991: 5).[8] In other words, negative experiences are not necessarily the reason why betrayal is bad. If you would betray someone knowing that this person would never find out anyway, most people intuitively still hold your actions to be bad. However, the problem with this line of reasoning is that betrayal is always bad *for someone*, and if the person is dead, there is nobody for whom the betrayal is bad. Elaborating Nagel's betrayal example, Fisher (1997) argues that the deprivation of good things in life can still be bad for an individual at the time when this individual is already dead: imagine a person's wife and best friend were in orbit in outer space and betrayed him. If the man happens to

---

[6] A being is a subject of its life when it has 'beliefs and desires; perception, memory, and a sense of the future, including their own future; an emotional life together with feelings of pleasure and pain; preference and welfare interests; the ability to initiate action in pursuit of their desires and goals; a psychophysical identity over time; and an individual welfare' (Regan, 2004 p. 243). Simmons, on the other hand, argues that to be able to experience suffering and pleasure already presupposes the characteristics on this list.

[7] Note that this is exactly why Epicure held that we should not be afraid of death: when you are dead you have no experiences, so you cannot suffer from death.

[8] Another question that the foregone opportunities account raises is whether abortion is a problem as well. After all, a fetus also loses its future opportunities for experiencing the goods in its life. However, Harman (2003) argues that this is not the case, because fetuses, at least until they experience pain and pleasure, do not have moral status. Whether or not they will have moral status in the future depends on whether or not their lives will be continued (up to the point where they can experience pain and pleasure). A consequence of this view is that once a fetus does acquire the ability to feel pain and pleasure, abortion is morally problematic. Another argument that has been put forward in this context is that the fetus is different from the person it will be in the future, so that the future opportunities for experiencing the goods in a person's life are simply non-existent in the foetal stage yet: there is no 'sufficient psychological continuity between the foetus and the lost future to say that the foetus *has been harmed* by the loss, though there may be an impersonal loss to the world' (Palmer, 2010: 137; italics in original).

be dead by the time light waves travel from outer space to earth, and in other words it will never be possible for him to experience anything bad as a result, in Fisher's view he would have still been harmed. Whether or not one finds examples such as these convincing seems to depend on one's intuition, however, and no consensus has as of yet been reached in this debate. However, we would like to point out one implication of rejecting the foregone opportunities account: if death were not a harm for animals due to their foregone opportunities of experiencing future goods, it would not be for human babies either (and in fact, if one goes along with our suggestion that the desire account is flawed, it may be difficult to argue that death is a moral harm for *any* humans).

14.5 Killing fish reflects badly on our character

So far, we have argued that the foregone opportunities-account of the harm of death is more convincing than the desire-account. However, the foregone opportunities-account rests on the assumption that one can be deprived of something even if one does not experience it as such, and whether or not one accepts the foregone opportunities-account in the end seems to depend on one's intuition regarding this quandary. However, even if we could argue that death were *not* a direct harm for fish, this may not automatically mean we are justified in killing them. One could argue on the basis of relational or virtue ethical accounts that we should not kill fish, either because this would not be virtuous and would reflect badly on our character or because of the responsibility we have towards fish in our care. How could such an argument proceed? Both of these accounts in animal ethics approach questions regarding our duties towards animals differently than the utilitarian or rights-based approaches discussed above. They do not start by asking the question whether and on what basis fish have moral status, but already assume that all sentient animals belong to our moral community (Bovenkerk and Meijboom, 2012). A virtue ethicist asks what is the right attitude to take towards animals and how our treatment of animals reflects on our character. According to some virtue-ethical accounts, human flourishing is foundational for our ethical aims in life (Walker, 2007). Acting virtuously is one of the constitutive elements of human flourishing, but we don't act virtuously just in order to achieve our own flourishing. As we are social beings, care for others' flourishing is part of what it is to be human. Animals can flourish in many of the same ways as humans and therefore if we have reason to care for the flourishing of other human beings, we have reason to care for that of animals (Walker, 2007).[9] A relational ethicist asks what obligations we have on the basis of the specific relationship we have with the animals in our care and the vulnerability of the animal in question (see for example Palmer,

---

[9] Aspects of flourishing that we share with animals are, for example, having an interests in 'a safe and comfortable place to sleep, enough to eat, appropriately satisfied sexual urges, sufficient room for exercise, clean water, sunshine (or darkness), appropriate social relations and hierarchy, physical health, positive psychological states' (Walker, 2007: 15).

2010; Swart and Keulartz, 2011). In the latter case, for example, this means that we have different obligations towards wild caught fish than towards farmed fish. Both accounts, then, focus not so much on the characteristics of the animals in question, but rather on the moral agent and her duties or her right attitude.

Combining a virtue ethics and a relational ethics account one could argue that when a person brings a fish into her 'circle' by either farming it, keeping it as a pet, or by performing experiments on it, one has the obligation to treat the fish with respect or to guarantee its welfare. If one does not do so, the person shows bad character; she takes the fish out of its own environment for her own purposes and treats it as if its flourishing does not matter. She would be committing the vices of callousness and *hubris.* She enters a relationship of exploitation with a vulnerable being. A question such an account would give rise to is what makes this person's behaviour callous and exploitative? What does it mean to fail to treat the fish with a lack of respect? Does killing it amount to a lack of respect even if killing were not directly harmful to fish? Virtue-ethical or relational accounts do not give unequivocal answers to these questions, and in that sense they seem to depend on other moral theories to be able to apply their views to specific cases. However, the start of a virtue- and relational ethical answer could be the following. Whether or not killing is a direct harm to a fish, we should not treat fish as if they were mere things, because then our attitude would show a lack of respect for and knowledge of other creatures' needs. Whether our attitude would show this lack of respect and knowledge to a large extend depends on the reasons for why a person kills fish. Is it out of pleasure of killing, because she wants to feel power over other living creatures, simply because she does not give the death of other creatures much thought, because she likes the taste of sushi, or is it out of necessity of feeding her family? The intention behind the action of killing makes an important difference for the evaluation of the character of the killer.

If we consider the vast numbers of fish that are routinely caught in the wild or raised on fish farms, the fact that fish in farms are referred to in terms of kilo's rather than number of individuals, and the fact that great numbers of fish are killed gratuitously as bycatch (see also 14.8), we cannot escape the feeling that in the process of production and consumption of fish many people do tend to view and treat them as if they were mere things. They are not mere things, however, but individuals with their own life that can go better or worse for them and with their own purposes. They are animals with a sophisticated biology that are capable of complex behaviour; some species of fish can change sex several times in their lifetime (Rodgers *et al.*, 2007), some can migrate home over thousands of kilometres to breed at the place they were themselves began their life (Odling-Smee and Braithwaite, 2003), some can cooperatively hunt (Bshary *et al.*, 2006) and some even play with other species of animal, even more bizarrely, some can move the position of their mouth within the course of 24 hours and climb a rock face (see also Driessen, 2013). Routinely killing fish does not reflect the respect that such features should give rise to. Fish may not have awareness or second-order intentions, but they are definitely not mere things either. Moreover, when contemplating whether it is a harm to kill a fish, we should also consider whether it has had a good and flourishing life, appropriate for the species to which it belongs. It may be more justified to kill a fish that has had a long and flourishing life than a fish that has lived a short and miserable life.[10] The foregoing has been rather speculative and needs to be further evolved.[11] Nevertheless, it seems at first sight at least that these arguments are strongest when directed to the routine commercial killing of fish and those situations where the killer's motives are merely pleasure seeking or borne out of ignorance.

14.6 How harmful is death for a fish?

Supposing that death is harmful for a fish, this still does not settle the question of whether we are allowed to kill fish for consumption or other purposes, such as experimentation or recreation. This depends on how this harm is to be weighed compared to other harms or benefits. When we make a utilitarian calculus it may be possible that the death of a number of fish would contribute to the overall good, for example if people need fish to survive. Even though this does not hold for people living in wealthy countries, for they have alternative sources of protein, it may hold for people in developing countries and, for example, Inuit who have no realistic alternatives at their disposal. Moreover, it has been argued that from an environmental perspective it is best overall when people eat fish, as this would contribute much less to climate change than other sources of protein (Kiessling, 2009, but see Röcklinsberg, 2012: 10 for a critical discussion of this viewpoint). Note, however, that it would be difficult to justify the vast numbers of fish being caught for consumption (including bycatch) today (estimated to be between 9.7×10^{11} and 2.7×10^{13} individuals (Mood and Brook, 2012), using a utilitarian calculus. Even one who argues from a rights-based position, could maintain that a right to life can be trumped, for example when the right to life of other beings is at stake; rights are not absolute. The main question then seems to be whether the right to life for fish, or the interests that fish have in continued living, counts the same as, or less, than the rights of humans or other animals? A widely shared intuition is that it is worse to kill a human being or another mammal than to kill a fish. But how can we account for this intuition?

Mary Anne Warren (1986) has argued that even though animals have a right to life, this right is weaker and can thus be overridden more readily than the human

---

[10] This is a quite liberal interpretation of Walker (2007).

[11] It should also be noted that some philosophers argue that we should not try to discover through rational argumentation whether killing animals wrongs them. We should simply listen to our emotions when we identify with animals, an identification which is based partly on the vulnerability we share with animals (cf. Diamond, 2008; Palmer, 2010; for a critical discussion of this view).

elements than the quality of life of humans. Our point, then, is not epistemological, but rather moral. It is not that we cannot know how animals experience certain sensations, but rather that we cannot judge whether one experience of one being is qualitatively better than a similar experience of another being, just as we cannot judge whether going to the opera is more valuable to us individually than going to a football match is for someone else. Yet, we should be careful not to make the fallacy of arguing from ignorance: after all, from the fact they we cannot judge whether one experience is better than another experience we cannot conclude that therefore both experiences are equally valuable.

According to this line of reasoning, it appears, then, that if we want to argue that most human lives are more valuable than most animal lives we would need to resort to an objective value theory that could show that regardless of the value of life as experienced by a particular being, certain goods and enjoyments are objectively superior to other ones. However, as we also briefly argued above, it is difficult to imagine what could be the basis of such an objective claim. Perhaps we could resort to a subjective value theory and argue that from the point of view of the animal or human in question it is worse to die the more conscious one is? Consciousness could then provide a common yardstick to measure the quality of life of different species (and within species) by. We could conceive of consciousness as a gradual notion; some animals have more and some less consciousness. Could those who have more consciousness experience the goods in their lives to a higher degree and could we therefore say that their life contains more value (Droege and Braithwaite, 2014)?[15]

14.7 Weaker interest does not mean no interest in continued life

Whether or not we could say that life has more value for humans than for other animals, from an unequal *value* of life we cannot automatically conclude an unequal *interest* in continued life, let alone draw specific action-guides.[16] If we were to consider that levels of consciousness or of creative and intellectual pleasures would make certain lives more valuable and that those with more valuable lives have a greater

---

[15] A different subjective view that would explain why death would be a lesser harm for animals is provided by Jeff McMahan (2002), who argues that when a being has less psychological connectedness it is deprived of less future goods if it is killed prematurely. According to his 'Time-relative interest account' we should discount welfare loss from being killed, relative to the lack of psychological connectedness in a being. McMahan's account has certain problematic implications, however (see Višak, 2015) and depends on a controversial view of personal identity.

[16] This point is analogous to the point made by Simmons (2011), who argues that from unequal quality of life we cannot conclude unequal *right* to life. After all, life does not have the same value for different human beings either, but we do not conclude from that fact that their right to life is not equally strong. Simmons proposes that everyone who meets a certain threshold of value should enjoy an equal right to life. If we are going to grant a right to life to marginal humans, we should also do so to animals that have similar or greater capacities for emotions and cognition. In this chapter, however, we focus not so much on a right to life, but only on the harm of death and the concurrent view that animals would have an interest in continued life.

interest in continued life, would we then not have to conclude that some humans are harmed more by death than other humans? And from the fact that some are harmed more by death, would we not have to conclude that some have a greater interest in continued life? Here again we can wonder whether there is a neutral vantage point to compare individuals' interests by. But even if there were, most people would not want to draw the conclusion that in a case where we would have to choose between saving a more intelligent or a less intelligent person, for example, we have to save the more intelligent one. This argument does not only refer to marginal humans, but also to normal, healthy adults, who also differ in their levels of consciousness and creative and intellectual powers. Even if we assume that (at least certain species of) fish have a lower quality of life than other animals, then, the question whether or not we are allowed to kill them is not settled yet. Still, our intuition remains that if we have to choose between killing a fish or killing a human or other mammal, we should kill the fish.

How can we explain this intuition? Either it could be understood as a predictable bias on the part of humans and our intuition is simply wrong. The intuition could simply be influenced by our culture or tradition. Or perhaps this intuition is based on the acknowledgement that self-conscious beings suffer psychologically from the idea that they will die. This, however, is not the type of harm that we are dealing with in this paper; we are addressing the question whether death *as such* is harmful for fish. After all, we think we should not kill humans even if they do not fear death or even when we would kill them painlessly in their sleep. Another alternative is to grant the option outlined above that those with a higher quality of life have a greater interest in continued life and that this should lead us to treat them differently. In that case we would also have to acknowledge that it is worse to kill normal adult humans than 'marginal humans' (including babies), and moreover, that it is worse to kill humans who can experience greater intellectual pleasures than humans with 'lower' intellectual pleasures. Yet, even if we would take this last, dubious avenue, this would still leave open the question of whether killing fish for consumption, recreation or animal experimentation is justified. Even if sentient animals and marginal humans would have a *weaker* interest in continued life, this does not mean they have *no* interest in it. In order to determine the justifiability of the use of fish, we would have to weigh different interests against each other.[17] Following Van de Veer (1979), we could distinguish basic from serious and peripheral interests and argue that basic interests should trump peripheral ones. We consider recreation to be a peripheral human interest, which does not trump fish basic interest in life. In the case of animal

---

[17] Analogously, Simmons (2009) suggests that we could still say that sentient animals have a right to life, even if it were a weaker right, and that this means that the right to life of humans can trump the right to life of fish in situations where these two rights conflict. However, in the case of killing fish for consumption there is, at least in rich countries, no conflict between the right to life of humans and fish and hence we cannot rely on the weaker right of life of fish to justify killing them.

experimentation according to this line of reasoning it may be allowed to kill fish if we can reasonably predict that this would save human lives. In the case of consumption, our judgment would have to rely on the specific situation at hand; in each case we would need to weigh how serious the human interest in fish consumption (and the revenue from fish production) is.

14.8 Implications

As we have seen, the implications of holding that death is harmful for fish need to be weighed in practice. In this section, we will discuss some implications of our views for three practices: wildlife management, sports fishing, and commercial fishing. One counterargument that has been brought forward to the claim that we are not justified in certain cases to kill fish is that we would need to stop predators from killing prey animals as well. So, if we should not kill fish, because death harms them, we should also stop sharks and dolphins from eating fish. This would lead to an absurd situation, where we end up creating parks for wild animals where we shelter, feed and protect them and where predators need to be fed on vegetarian diets. A common response to this charge is that while *we* are not allowed to kill animals since we can contemplate the consequences of our actions and act morally, we cannot expect this from animals that are not moral actors. As Cohen (Cohen and Regan, 2001) has pointed out, however, if we should not stop a predator from killing a prey because it is not a moral actor, we should not stop a predator from killing a human infant either, and this is counterintuitive. Whether or not the predator can be held morally responsible for its actions is irrelevant to the question of whether we should intervene if we can.

Simmons (2009) gives a more convincing argument. We should make a distinction between negative and positive duties to others. Negative duties tell us to refrain from acting in certain ways and positive duties tell us to perform specific actions. More specifically, negative duties tell us to refrain from harming others, whereas positive duties tell us to undertake steps to ensure that others are benefitted. Negative duties have a more overriding character than positive ones. When we want to decide whether we have a positive duty to help someone, we need to take into account other circumstances of the case at hand as well. In this particular instance, we should take into account the effects of our actions on ecosystems. Interfering in predator-prey relationships would likely have grave ecological consequences. Often, we simply do not know what the consequences would be, and therefore we should interfere as little as possible. This is different when humans are concerned because human society and wild nature are separated to the extent that saving humans from predators would not have bad ecological effects. After all, no predators are dependent on humans as their main source of food. The fact that we do not have a duty to rescue prey animals does not mean that they have no right to life, only that our positive duty to contribute to that life is outweighed by other factors.

A practice for which the view that death is a harm for fish also has implications, is that of recreational angling. We should not only weigh the harm done to the fish against the pleasure that humans derive from angling, but our views also have implications for the way in which angling is conducted. Recreational angling is an extremely popular pastime. Estimates have proposed that as many as 47.1 billion fish are caught this way annually (Cooke and Cowx, 2004), but not all these fish are killed. As many as two thirds of them are caught and then released back into the water body they came from. Recreational anglers in many different countries practice this 'catch and release' system of fishing (Bartholomew and Bohnsack, 2005). Sometimes this is because the fish they catch are below a certain minimum size and regulations require sub-legal sized fish to be released. Elsewhere, the increasing popularity of recreational fishing has led to issues with too many anglers, but too few fish, so in the interests of the conservation of the fish in the system, the anglers will release fish that they catch so that they are not extirpated from a specific lake or river. There are some places where only catch and release fishing is permitted (Arlinghaus *et al.*, 2007).

Different angling organizations have developed educational materials regarding the best practice for catch and release. These describe the kinds of fishing gear that should be used, how the fish should be handled (minimizing the time the fish is brought out of water, for example), and being aware that angling during very warm weather can increase the mortality of the fish after they are released (Cooke and Suski, 2005). There are also guides that explain when fish should be killed rather than released because sometimes injuries caused by fishing hooks, or barometric trauma (experienced when a fish is brought to the surface too quickly from a depth of 12 m or more) may be too devastating to support post-release survival. Fish that become hooked through their gills or with the hook caught in their gut have much lower survival rates if they are released.

Thus the practice of catch and release raises several concerns. If the fish is severely wounded and will ultimately die then putting it back into the water after it has been unhooked, or if the hook cannot be removed after the line is cut, will lead to a slow death. Better practice surely, in these cases, would be to kill the fish quickly while it is still captive. Another related issue is the stress associated with being caught. If recaptures happen multiple times over several days, there is a strong chance the fish will become chronically stressed which can alter the stress physiology of the fish such that the fish becomes immuno-compromised (Barton, 2002). Under such a scenario, there is an increased chance that the wound where the hook pierced the fish's skin may become infected, or the overall capacity for the fish to cope with future capturing and handling, or other environmental challenges such as the threat of predation may be impaired leading to a higher risk of mortality.

Relatively recent changes in legislation in some countries (such as Germany and Switzerland) now require that fish that are not protected by certain size limits should be killed when caught (Arlinghaus *et al.*, 2007). This is because the chance that the fish will experience prolonged suffering is regarded to make the practice of catch and release unethical. Such a viewpoint has proved to be contentious, with some arguing that if fish are handled appropriately then catch and release is not an ethical issue, whereas mandates that call for fish above a legal threshold to be killed after capture could be an environmental ethics problem because it may result in compromised conservation for certain fish populations (Arlinghaus *et al.*, 2007). Is it possible to ensure proper handling and care is taken during the catch and release process? Such questions are hard to ignore when anglers fishing in very popular locations catch fish with multiple scars from previous encounters with hooks (Policansky, 2007).

A related issue is the problem of bycatch: the vast numbers of non-target species of sea animals that are caught 'accidentally' during large-scale commercial fishing operations. The use of high tech fishing methods, such as longlines, trawling, and purse seining, makes it difficult, if not impossible, to select only the target species (Gilman *et al.*, 2006). It is not uncommon that 80 to 90% of captured animals are thrown overboard in order to keep only the commercially viable species, such as shrimp and tuna.[18] It has been estimated that globally 40% of all caught sea creatures is discarded, amounting to 63 billion pounds annually (Keledjian *et al.*, 2014). Also endangered and protected species, such as turtles and octopuses, are regularly thrown overboard as bycatch. Most animals that are thrown overboard die as a result of stress or injuries, if they did not already die as a result of catching method or of being left to suffocate on deck. Bycatch is regarded as one of the main threats to the maintenance of healthy populations of fish and flourishing marine ecosystems.[19] Debate is ongoing on the rules stipulating what fishermen should do with bycatch: which species and how many fish should be returned to the sea and how can harm to these animals be minimized?[20] Another issue that is debated is whether numbers and typical species of bycatch should be labelled on fish products (Foer, 2009).

What seems to be missing from these debates is the idea that death in itself could be harmful for fish. The concerns that are taken into consideration are firstly, whether the fish suffer from their wounds and perhaps die a painful death as a result, and secondly, what the consequences of killing versus catch and release or of bycatch are for the conservation of fish populations. It appears to be taken for granted that killing fish is

---

[18] See http://tinyurl.com/px7jspy.

[19] See http://tinyurl.com/orb4nuz.

[20] http://tinyurl.com/oc2rylz. Ironically, some argue that bycatch should not be reduced, because it provides food for seabirds that would otherwise prey on smaller seabirds which are endangered. This argument bypasses the problem that throwing bycatch overboard has led to an increase in the numbers of large seabirds in the first place. See http://tinyurl.com/ob4sqsg.

only problematic when the fish suffer in the process, and that it is better to kill a fish than to have it live on and suffer. In other words, only suffering, and not dying as such, is held to be morally relevant. While it may be better to kill a fish than leave it to suffer when extreme suffering is concerned, the view that death itself is harmful for fish, complicates this conclusion. If death is a harm for fish because it forecloses valuable opportunities in their lives, then perhaps a certain degree of suffering is justified if there is a good chance that the fish survive. Another tacit assumption in this debate seems to be that the death of fish is not problematic for individual fish but only for the population to which they belong. This view could either be based on anthropocentric or on ecocentric concerns, but seems to overlook zoocentric concerns, dealing with the moral status of individual sentient beings. Taking the interest of individual fish in continued life into account would mean that angling may still be morally problematic, even if fish don't suffer as a result (as many anglers claim) (Arlinghaus *et al.*, 2007) or if wild fish stocks remain healthy. Similarly, bycatch would be problematic in itself, and not only because of the suffering of the fish involved, but also because of its effect on the health of fish populations or ecosystems. Note that this issue of the gratuitous killing of large numbers of 'useless' individuals finds parallels in for example the problem of one-day old male chicks that are killed as they are superfluous to egg production (see Aerts and De Tavernier, 2016).

14.9 Conclusions

In this chapter, we have examined several arguments for the claim that death is a harm in a morally relevant sense and we have applied these to the case of fish. It is unlikely that fish are self-conscious or have categorical desires. However, these capacities are relevant only for a desire-account of the harm of death and in our view this account is flawed. A more promising take on the harm of death is offered by the foregone opportunities account, which argues that death is a harm because life is instrumentally valuable for all beings that can have experiential well-being. This would mean that sentience is a sufficient capacity to be able to be harmed by death and fish are (most likely) sentient. Nevertheless, the foregone opportunities-account rests on the assumption that one can be deprived of something even if one does not experience it as such, and this assumption is the subject of a complicated on-going philosophical debate. Still, even if we could argue that death were *not* a direct harm for fish, this does not automatically mean that we are justified in killing them.

We have briefly explored whether a virtue-ethical argument, combined with a relational argument could show that we should not kill fish, because this would reflect badly on our character, especially when it concerns fish in our care. These arguments are stronger when directed to the routine large-scale killing of fish than in the case of one particular fish. Next, we have examined how one could account for the commonly shared intuition that it is worse to kill a human being than to kill a fish. Some argue that even if we apply the equality of interest-principle, it is still possible to differentiate between different species, because members of some species have a higher interest in continued life than others. However, even if we could argue that some lives contain greater value than others, this does not necessarily mean that some beings have a greater interest in continued life than others, nor does it settle the question of whether fish may be killed. We have suggested that the latter depends on a weighing of basic, serious, and peripheral interests of humans in killing fish for consumption, recreation, or experimentation against the basic interest of fish in survival. Moreover, it should be noted that even though we have focused here on the question of whether painless killing is a harm in itself, in reality very rarely killing is not accompanied by pain and suffering. In order to answer the question whether we may kill fish in a particular case, the suffering inflicted should of course be taken into the equation as well. Finally, we have discussed the implications of our arguments for three debates, the catch and release versus the catch and kill debate, the debate surrounding bycatch, and the debate about whether we should stop predators from killing prey.

References

Aerts, S. and De Tavernier, J., 2016. Killing animals as a matter of collateral damage. In: Meijboom, F.L.B and Stassen, E.N. (eds.) The end of animal life: a start for ethical debate. Wageningen Academic Publishers, Wageningen, the Netherlands, pp. 169-185.

Arlinghaus, R., Cooke, S.J., Lyman, J., Policansky, D., Scwab, A., Suski, C., Sutton, A.G. and Thorstad, E.B., 2007. Understanding the complexity of catch-and-release in recreational fishing: an integrative synthesis of global knowledge from historical, ethical, social, and biological perspectives. Reviews in Fisheries Science 1: 75-167.

Bartholomew, A. and Bohnsack, J.A., 2005. A review of catch-and-release angling mortality with implications for no-take reserves. Reviews in Fish Biology and Fisheries 15: 129-154.

Barton, BA., 2002. Stress in fishes: a diversity of responses with particular reference to changes in circulating corticosteroids. Integrative and Comparative Biology 42: 517-525.

Bovenkerk, B. and Meijboom, F.L.B., 2012. The moral status of fish: the importance and limitations of a fundamental discussion for practical ethical questions in fish farming. Journal of Agricultural and Environmental Ethics 25: 843-860.

Bracke, M.B.M., 1990. Killing animals, or, why no wrong is done to an animal when killed painlessly. University Utrecht, Utrecht, the Netherlands.

Braithwaite, V. and De Perera, T.B., 2006. Short-range orientation in fish: how fish map space. Marine and Fresh Water Behaviour and Physiology 39: 37-47.

Braithwaite, V., 2010. Do fish feel pain? Oxford University Press, New York, NY, USA, 256 pp.

Braithwaite, V.A. and Huntingford, F.A., 2004. Fish and welfare: do fish have the capacity for pain perception and suffering? Animal Welfare 13: 87-92.

Bshary, R.,A. Hohner, K., Ait-el-Djoudi and Fricke, H., 2006. Interspecific communicative and coordinated hunting between groupers and giant moray eels in the red sea. PLoS Biology 4 (12): e431.

Cigman, R., 1981. Death, Misfortune and Species Inequality. Philosophy & Public Affairs 10(1): 47-64.

Cohen, C. and Regan, T., 2001. The animal rights debate. Rowman & Littlefield Publishers, Lanham, MD, USA, 336 pp.

Cooke, S.J. and Cowx, I.G., 2004. The role of recreational fisheries in global fish crises. BioScience 54: 857-859.

Cooke, S.J. and Suski, C.D., 2005. Do we need species-specific guidelines for catch-and-release recreational angling to conserve diverse fishery resources? Biodiversity and Conservation 14: 1195-1209.

Davidson, D., 1982. Rational animals. Dialectica 36: 318-327.

DeGrazia, D., 1996. Taking animals seriously: mental life and moral status. Cambridge University Press, Cambridge, UK.

DeGrazia, D., 2002. Animal rights: a very short introduction. Oxford University Press, New York, NY, USA.

DeGrazia, D., 2009. Self-awareness in animals. In: Lurz, R.W. (ed.) The philosophy of animal minds. Cambridge University Press, Cambridge, UK.

Diamond, C., 2008. The difficulty of reality and the difficulty of philosophy. In: Cavell, S., Diamond, C., McDowell, J., Hawking, I. and Wolfe, G. (eds.) Philosophy and animal life. Columbia University Press, New York, NY, USA, pp. 43-89.

Driessen, C., 2013. In awe of fish? Exploring animal ethics for non-cuddly species. In: Röcklinsberg, H. and Sandin, P. (eds.) The ethics of consumption. The citizen, the market and the law. Wageningen Academic Publishers, Wageningen, the Netherlands, pp. 251-256.

Droege, P. and Braithwaite, V.A., 2014. A framework for investigating animal consciousness. In: Lee, G, Illes, J. and OHL, F. (eds.) Ethical issues in behavioral neuroscience. Current Topics in Behavioral Neurosciences 19, Springer-Verlag, Berlin, Germany, pp. 79-98.

Duncan, I.J.H., 2006. The changing concept of animal sentience. Applied Animal Behaviour Science 100: 11-19.

Ebbesson, L.O.E. and Braithwaite, V.A., 2012. Environmental impacts on fish neural plasticity and cognition. Journal of Fish Biology 81: 2151-2174.

Fischer, J.M., 1997. Death, badness, and the impossibility of experience. Journal of Ethics 1: 341-353.

Foer, J.S., 2009. Eating animals. Hamish Hamilton, London, UK.

Francione, G., 2006. Equal consideration and the interest of animals in continuing existence: a response to professor Sunstein. University of Chicago Legal Forum 231: 239-240.

Gilman, E.L., Dalzell, P. and Martin, S., 2006. Fleet communication to abate fisheries bycatch. Marine Policy 30(4): 360-366.

Harman, E., 2003. The potentiality problem. philosophical studies. International Journal for Philosophy in the Analytic Tradition 114(1/2): 173-198.

Huntingford, F.A., Adams, C.E., Braithwaite, V.A., Kadri, S., Pottinger, T.G., Sandoe, P. and Turnbull, J.F., 2007. Current understanding on fish welfare. Journal of Fish Biology 70: 1311-1316.

Kaldewaij, F., 2006. Animals and the harm of death. In: Kaiser, M. and Lien, M. (eds.) Ethics and the politics of food. Wageningen Academic Publishers, Wageningen, the Netherlands, pp. 528-532.

Keledjian, A.,Brogan, G., Lowell, B., Warrenchuk, J., Enticknap, B, Shester, G., Hirshfield, M. and Cano-Stocco, D., 2014. Wasted catch: unsolved problems in U.S fisheries. Available at: http://tinyurl.com/kore6hg.

Kiessling, A., 2009. Feed – The key to sustainable fish farming. Fisheries, sustainability and development. Royal Swedish Academy of Agriculture and Forestry (KSLA): 303-322.

Lambooij, B., Kloosterboer, K., Gerritzen, M.A., André, G., Veldman, M. and Van de Vis, H., 2006. Electrical stunning followed by decapitation or chilling of African catfish (*Clarias gariepinus*): assessment of behavioural and neural parameters and product quality. Aquaculture Research 37: 61-70.

Lurz, R.W. 2009. The philosophy of animal minds. Cambridge University Press, Cambridge, UK.

McCloskey, H.J., 1979. Moral rights and animals. Inquiry 22: 23-54.

McMahan, J., 2002. The ethics of killing. Problems at the margins of life. Oxford University Press, Oxford, UK.

Mill, J.S., 1957. Utilitarianism. The Bobbs-Merrill Company Inc., New York, NY, USA (originally published 1861).

Mood, A. and Brook, P., 2012. Estimating the number of farmed fish killed in global aquaculture each year. Fishcount, UK. Available at: http://tinyurl.com/qxao6o7.

Nagel, T., 1991. Mortal questions. Cambridge University Press, Cambridge, UK.

Nilsson, J.T.S., Kristiansen, J.E., Fosseidengen, A., Fernö and Van den Bos, R., 2008. Learning in cod (*Gadus Morhua*): long trace interval retention. Animal Cognition 11: 215-222.

Odling-Smee, L. and Braithwaite, V.A., 2003. The role of learning in fish orientation. Fish & Fisheries 4: 235-246.

Palmer, C., 2010. Animal ethics in context. Columbia University Press, New York, NY, USA.

Pluhar, E.B., 1995. Beyond prejudice: the moral significance of human and nonhuman animals. Duke University Press Books, Durham, NC, USA, 395 pp.

Policansky, D., 2007. The good, bad and truly ugly of catch and release: what have we learned? Wild Trout Symposium 9: 195-201.

Regan, T., 2004. The case for animal rights. University of California Press, Berkeley, CA, USA.

Röcklinsberg, H., 2012. Fish for food in a challenged climate: ethical reflections. In: Potthast, T. and Meisch, S. (eds.) Climate change and sustainable development. Ethical perspectives on land use and food production. Wageningen Academic Publishers, Wageningen, the Netherlands, pp. 326-334.

Rodgers, E.W., Earley, R.L. and Grober, M.S., 2007. Social status determines sexual phenotype in the bi-directional sex changing bluebanded goby *Lythrypnus dalli*. Journal of Fish Biology 70: 1660-1668.

Silverstein, H.S., 1980. The evil of death. Journal of Philosophy 77: 414-415.

Simmons, A., 2009. Animals, predators, the right to life, and the duty to save lives. Ethics & the Environment 14: 15-27.

Simmons, A., 2011. Do all subjects of a life have an equal right to life: the challenge of the comparative value of life.

Singer, P., 1980. Animals and the value of life. In: Regan, T. (ed.) Matters of life and death. Random House, New York, NY, USA.

Singer, P., 2011. Practical ethics. Cambridge University Press, Cambridge, UK, 395 pp.

Sneddon, L.U., Braithwaite V.A. and Gentle, M.J., 2003. Novel object test: examining nociception and fear in the rainbow trout. Journal of Pain 4: 431-440.

Swart, J. and Keulartz, J., 2011. Wild animals in our backyard. A contextual approach to the intrinsic value of animals. Acta Biotheoretica 59: 185-200.

Tooley, M., 1972. Abortion and infanticide. Philosophy & Public Affairs 2: 37-65.

Van De Vis, H., Kestin, S., Robb, D., Oehlenschläger, J., Lambooij, B., Münkner, W., Kuhlmann, H., Kloosterboer, K., Tejada, M., Huidobro, A., Otterå, H., Roth, B., Sørensen, N.L., Akse, L, Byrne, H. and Nesvadba, P., 2003. Is humane slaughter of fish possible for industry? Aquaculture Research 34: 211-220.

VandeVeer, D., 1979. Interpecific justice. Inquiry 22: 55-70.

Višak, T., 2011. Killing happy animals: explorations in utilitarian ethics. Quaestiones Infinitae 66.

Višak, T., 2015. Environmental Ethics. In: Vedwan, N. (ed.) An integrated approach to environmental management. Wiley, Hoboken, NJ, USA, pp. 337-361.

Walker, R.L., 2007. The good life for non-human animals: what virtue requires of humans. In: Walker, R.L. and Ivanhoe, P.J. (eds.) Working virtue: virtue ethics and contemporary moral problems. Oxford University Press, New York, NY, USA, pp. 173-189.

Warren, M.A., 1986. Difficulties with the strong animal rights position. Between the Species 2(4): 433-441

15. Will wild make a moral difference?

B. Gremmen
Wageningen University, Hollandseweg 1, 6706 KN, Wageningen, the Netherlands;
bart.gremmen@wur.nl

Abstract

This chapter is about the ethics of killing wild animals. What is our moral reference while killing wild animals? Can we use norms of killing domesticated animals when we kill wild animals? Is the life and death of wild animals out of our moral reach by definition? Do we respect the wildness of an animal? Are there situations in which humans have to kill wild animals? In these cases humans are confronted with wild animals and the question can be asked: do we have to kill them? Our approach is to reformulate the three core questions of this book to the situation of wild animals. In answering the first question, what concepts are needed for the public and ethical evaluation of killing wild animals, we describe wildness as a broad concept, and equate it with parts of nature that are not controlled by humans. Our perspective on wildness is to consider it as a quality in specific individual animals, of being wild or un-wild. We differentiate between nine categories of animals in natural areas, and wild animals are considered to be at one end of a continuum and domesticated animals are at the other end. Thus, a development is possible from the wild stage to the pseudo-domestication stage and back again to the semi-wild stage. The description of an ethical framework of three principles enabled our affirmative answer to the second question, is it possible to justify the killing of a wild animal, and if so under what conditions. When we apply the ethical framework to the killing of wild animals, de-domesticated and feral animals, and to the killing of animals in pest control, the answer to the third question: 'Can we legitimately differentiate the issue of killing wild animals in different wild animal contexts, leads to seven conclusions?'

Keywords: eco-ethics, mid-level principles, de-domesticated and feral animals, pest control

15.1 Introduction

On December 12th 2012 a humpback whale beached on the shoreline of 'De razende bol', a small unhabituated island between the island of Texel and the city of Den Helder in the Netherlands. The whale was stuck on the beach and could not return to the water on its own. Not so long ago humans living nearby would have killed the animal immediately and its remains would have been used for all kinds of purposes. In our modern times we try to save its life. In only a few days' time this animal, named Johannes, became a national symbol of helping a wild animal in need. Political parties,

*Franck L.B. Meijboom and Elsbeth N. Stassen (ed.) **The end of animal life: a start for ethical debate***
DOI 10.3920/978-90-8686-808-7_15, © Wageningen Academic Publishers 2016

civil servants, scientists, and members of societal organizations were not only engaged in debates, but also in rescue and euthanasia attempts. All these attempts failed and the whale eventually died. In the remaining debate ecologists and nature conservationists still argued not to kill dying wild animals in distress, while the majority of the other participants in the debate argued for a humane death of these animals.

At first sight it seems humans need not to be involved at all when wild animals die: they are wild because they take care of themselves in areas where they are out of control of humans. In those situations wild animals die because of very different reasons: hunger, thirst, disease, predators, and also of the consequences of old age. Humans often are unaware of the fact that, out of their sight, animals could be dying. However, sometimes people are confronted with dying wild animals, like the beached humpback whale in 2012. This chapter is about the ethics of killing wild animals. What is our moral reference while killing wild animals? Can we use norms of killing domesticated animals when we kill wild animals? Is the life and death of wild animals out of our moral reach by definition? Do we respect the wildness of an animal or do we even take away the wild status of a humpback whale? Are there situations in which humans have to kill wild animals? These questions pertain not only to killing, but also about who and when, and in what way we must kill wild animals. Although in the past most animals killed by humans were killed by hunters, the practice of hunting as such is left out of the analysis.

It seems that in order to answer these questions we can rely on animal ethics: the moral framework of the killing of domesticated animals. According to Dutch law, based on this theory, humans are obliged to help an animal in distress. From that ethical perspective the first duty of humans is to save or help individual wild animals in situations where humans are present. When all help fails our second duty, if possible, is to kill these animals in a humane way. The example of the humpback whale seems to fit into this scheme because it is an individual animal surrounded by humans. However, from the perspective of eco-ethics wild animals are part of ecosystems. Therefore the focus is on groups and species rather than on individual animals. In general this ethical framework advocates respect for the wildness of animals. In cases of dying wild animals, like the humpback whale, the eco-ethics ethical framework advices a hands off strategy. This seems to lead to a stale mate between two rivalling ethical frameworks, and thus leaving nature management caught between two sets of norms governing animals and nature.

If we regard some ethological distance of the dualism between 'wild' and 'tame', all kinds of intermediary shades appear. Also the number of situations in which humans have to decide to kill wild animals increases considerably. The humpback whale is an example of an individual wild animal in distress. We may consider this situation as bad luck and exceptional. But what about lost or abandoned seal babies on the shore lines of the Netherlands, Germany and Denmark? When we locate these animals, do we help them by bringing them to a shelter? Do we have to kill them on the spot or leave them alone to die? Other examples are weak or dying animals in nature parks like the Oostvaardersplassen in the Netherlands (Gremmen, 2014), and exotic animals that are destroying the biodiversity of an area. Also animal pest management is an example. In these cases humans are confronted with wild animals and the question can be asked: do we have to kill them? Our approach is to reformulate the three core questions of this book to the situation of wild animals:

1. What concepts are needed for the public and ethical evaluation of killing wild animals?
2. Is it possible to justify the killing of a wild animal, if so under what conditions?
3. Can we legitimately differentiate the issue of killing wild animals in different wild animal practices and contexts?

15.2 All kinds of wild animals

In this paragraph we discuss the concepts that are needed for the public and ethical evaluation of killing wild animals. What do we mean by 'wild' animals? Wildness may be conceived as a broad concept and equated with parts of nature that are not controlled by humans. This would mean that the animals are autonomous (Evanoff, 2005) in the sense that they are wholly free of deliberate human intervention. To protect this kind of nature, two recognizable trends follow the more traditional approach to wilderness preservation: trying to look after whatever is left of original nature, and the more recent ecological restoration approach, which 'initiates or accelerates the recovery of an ecosystem with respect to its health, integrity and sustainability' (Staatsbosbeheer, 2012). Ecological restoration consists of reforestation, lake restoration, elimination of non-native species. It seems impossible to preserve this kind of wildness in human-run, artificial environments (Jamieson, 1995) because difficulties arise with feral animals and also with de-domestication in landscapes heavily influenced by humans.

Our perspective on wildness is to consider it as a quality in specific individual animals, of being wild or un-wild. In this way also feral and de-domesticated animals are wild. Koene and Gremmen (2002: 130) differentiate between nine categories of animals in natural areas:

1. domesticated animals;
2. escaped domesticated animals, not able to cope independently;
3. released domesticated animals, not able to cope independently;
4. escaped animals, able to cope independently, but without natural behaviour;
5. released animals, able to cope independently, but without natural behaviour;
6. escaped animals, able to cope independently, and behaving in a natural way (feral animals);

7. released animals, able to cope independently, and behaving in a natural way (de-domesticated animals);
8. released animals, able to cope independently, and behaving in manner designed by humans (pseudo de-domesticated animals);
9. wild animals.

From a developmental perspective these categories presuppose a first period of only wild animals and a second period of wild and domesticated animals. After these periods a return to a kind of wild stage is possible: from the stage of domestication, via the feral stage, towards the de-domestication stage, and finally the wild stage (Koene and Gremmen, 2002: 130). De-domestication can be viewed as an end in itself: as a sort of species restoration, a way of getting populations of animals to resemble their wild ancestors not only in appearance but also in terms of behaviour. But it is most often advocated as means to an end: as part of a complex process of ecological restoration aiming to increase the so-called wildness and naturalness of an area in a long-term nature management strategy (Vera, 2009). What makes de-domestication different from other forms of nature restoration is that it involves deliberate intervention at the genetic level as well as conventional landscape management. Many generations are considered necessary to accomplish real (or as real as possible) de-domestication through changes in genotype at population level. The process of de-domestication initially involves the development of distinct, more fully adapted behaviour (in terms of natural group formation, leadership and rutting period, and so on) and selection pressure to initiate genetic changes over generations. De-domestication, as here described, shares certain characteristics with ecological restoration: suitable reference points must be found, valid data must be used to flesh the scheme out, and the present state of the environment must be compared with the conditions prevailing when the environment originally existed (Gamborg *et al.*, 2012). De-domestication is therefore an exercise in approximation with an unpredictable result and an end-point that is hard to define. Is populating the landscape with animals through de-domestication a second-rate imitation of the real thing, fake nature (Elliot, 1982), or wildness by proxy (Gremmen, 2014)? In the mainstream literature on animal welfare and ethics it is difficult to find clear, unanimous answers to this question. In the next paragraph our starting point will be the familiar distinction between animal welfare and our duties to eco-systems (Gamborg *et al.*, 2012).

15.3 An ethical framework of three 'mid-level' principles

Is it possible to justify the killing of wild, feral or de-domesticated animals, if so under what conditions? To answer these questions we could, like many authors in the field, try to apply a single, broad ethical theory about our duties to wild animals. In this paragraph two of these theories, animal welfare ethics and eco-ethics, will be discussed. However, even if we would find a suitable broad ethical theory, there would be no direct uncontroverted passage from theory to practice. According to Beauchamp and Childress (2001: 405), 'Concepts are too general, principles too indefinite, and the facts of cases too difficult to bring under principles. Even if we had a theory to supply our initial norms, we have no direct way to move from the theory to decisions in particular cases'. One of these authors concludes that 'truly practical judgments cannot be squeezed from abstract principles and general ethical theories alone' (Beauchamp, 2003: 12). The approach of Beauchamp and Childress to bioethics provides a pluralist approach that does not allow any one set of high level principles to have guaranteed priority in all ethical decision-making. Instead, four 'mid-level' principles are defended: respect for autonomy, non-maleficence, beneficence, and justice are each claimed to establish prima facie (on first appearance) obligations (Beauchamp and Childress, 2001). These four 'mid-level' principles are derived from general ethical theories, such as utilitarianism and Kantian ethics. These principles have been the inspiration and basis of the 'Ethical matrix' (Mepham, 1996), a tool for analysing ethical issues, but have also been the inspiration of the three 'mid-level' principles Eggelton *et al.* (2003) derived from the two broad ethical theories, animal welfare ethics and eco-ethics. In the next section we will use these three principles as an ethical framework to justify the killing of wild, feral or de-domesticated animals in different contexts.

The first broad theoretical approach is animal welfare ethics which focuses on individual animal welfare. Particularly in Europe a perceived need emerged for initiatives which placed limits on the use of animals for purposes to which most people agree (Gamborg *et al.*, 2012). Such initiatives fall under the heading of 'animal welfare' and their ethical ramifications are discussed in animal ethics. Although this approach has been developed in the context of pets and production animals, it is also applied to wild animals. Within animal ethics there are two main perspectives: a utilitarian view focuses on sentient animals, and a deontological view focuses on animal rights.

From a utilitarian view it is argued that sentient animals should be granted direct moral consideration (Singer, 1990), because the well-being of sentient animals matters to them. This means that they should not be deliberately harmed (except in situations of necessity, e.g. for self-preservation). Hickling (1994) emphasizes, however, that the philosophical theories justifying this ethical concern do not appear to allow any morally significant distinctions to be made between native and exotic species, or to attribute any overriding moral significance to the protection of unique indigenous species and ecosystems. For example, utilitarian-based ethical concern focused on the maximization of aggregate utility (whether defined in terms of pleasure/pain or preference satisfaction) does not support the harming of large numbers of sentient exotic mammalian pest species deemed necessary to protect small numbers of rare native species. According to the utilitarian view, what matters in our dealings with animals is our impact on their well-being. We should always aim to act so as to achieve the largest total sum of well-being (Singer, 1990). From a utilitarian view there is no clear divide in principle between the way we are required to treat domestic animals and the way we are required to treat wild animals. As effectively as possible, we should look after the well-being of animals, whether domestic or wild. The only difference is a pragmatic one: in practice, it may be more difficult to look after the well-being of wild animals than it is to look after the well-being of animals in our direct care. The arguments of utilitarian animal ethics lead to the first 'mid-level' principle: *take the interests of animals seriously into ethical consideration and avoid unnecessary harm to sentient creatures* (this principle applies to domestic, captive and wild species, irrespective of native or introduced status) (Eggleston *et al.*, 2003).

The second view, the strict animal rights or liberation approach, believes animals must not be exploited regardless of the benefits to humans or the environment (Regan, 1983). According to animal rights advocates like Regan (1983) many animals, including all vertebrates, have an inherent value of their own, based on their nature and capacities. They are not to be treated as instruments for someone else's use and benefit. Inherent value cannot be traded off, factored into calculations about consequences, or replaced. Creatures that possess it have basic moral rights, including the right to life and to liberty. On this view animal production should simply be stopped because it is bound to violate animal rights. The broad concept of a prima facie moral duty of non-interference in the lives of autonomous living creatures (including introduced wild species), argued for by animal rights theorists such as Regan, should be taken seriously. Such a duty can be argued for simply on the basis of acknowledgment of the real possibility that some non-human lives may have value in their own right, and not just insofar as they are useful or pleasing to humans (Brophy, 1972; Rollin, 1996). The arguments of the deontological rights ethics in animal ethics lead to the second 'mid-level' principle: *avoid interference in the lives of wild creatures. If intervention is unavoidable, it should at least be minimized* (Eggleston *et al.*, 2003). Individuals and groups have the right to pursue their particular interests provided these interests are not detrimental to reasonable conservation objectives or harmful to (morally significant) others.

The second broad theoretical approach is the bio-centric or eco-centric ethic, based on the inherent or intrinsic value of the environment and its constituents (species, populations, or ecosystems). Many of our beliefs and activities in wildlife management are aimed at preserving the environment, as expressed in Leopold's (1949) land ethic: 'A thing is right when it tends to preserve the integrity, stability, and beauty of the biotic community. It is wrong when it tends otherwise'. Eco-centric ethics has been developed more recently by philosophers such as Callicott (1980, 1987, 1989) and Rolston (1988, 1994). An eco-centric understanding of ethical responsibility places greatest moral significance on the protection of the integrity or health of whole ecosystems (and of viable populations of the species contributing to that integrity or health). The arguments from eco-ethics lead to the third 'mid-level' principle: *human-introduced wild mammal species proven to cause irreversible harm to the capacity for natural autonomy and wildness of lands and vulnerable native species should be managed where effective means are available* (Eggleston *et al.*, 2003). Interventions should not be so intensive and sustained as to result in an even greater loss to natural autonomy/wildness than would follow non-intervention (Eggleston, 2002). The conflict between the welfare of individual sentient animals (animal ethic) and concern for the ecosystem as a whole (bio-centric ethic) has formed the core conflict in the management of wild animals. However, the third 'mid-level' principle allows for a kind of pluralism in which different standards exist for, on the one hand, dealing with feral and de-domesticated animals and, on the other hand, wild animals, as suggested by Klaver *et al.* (2002).

15.4 Killing wild animals in different contexts

Can we legitimately differentiate the issue of killing wild animals in different wild animal contexts? We will use the ethical framework of the treatment of wild animals to consider all relevant values and concerns in three contexts in which wild animals are killed: wild animals, de-domesticated animals, and pest control. In each of these three contexts we will formulate conclusions about the killing of wild animals. If two or more of the principles of our ethical framework have conflicting implications, ethical consideration may identify one as resulting in the more pressing duty, and hence overriding the others (Eggleston *et al.*, 2003).

15.4.1 Wild animals

In the past wild animals were used as hunting animals, or as game, or were killed as pests or vermin. Efforts are being made all over the world to not to kill wild animals and their habitats. Their natural autonomy rather than simply untrammelled nature is valued (Hettinger and Throop, 1999). We may not kill wild animals for our own purposes; and as a consequence the main management strategy concerning wild animals is a 'hands-off' recommendation. This is most apt for relatively unmodified lands and wild co-evolved native species, and is not applicable in any blanket fashion in situations already subject to human involvement (Regan, 1983). Here, humans do not use animals in an instrumental way or are active in bringing the relevant animals into the world, nor is there any direct involvement in their upbringing. Thus, humans have less direct responsibility than we have for domestic animals. This leads to the *first (general) conclusion* about the killing of wild animals: *humans are not allowed to kill wild animals within their historical ecological and co-evolved biotic communities.* It is based on the (negative) right to non-interference for wild creatures (the second principle). This may explain, in the case of the beached humpback whale, the position of some people who argued not only against helping the animal, but also against killing it. However, according to Dutch law people must also help wild animals in need. This may be defended from the duty we have to avoid unnecessary harm to sentient creatures (the first principle). In many situations wild animals are in trouble because of human influence on the environment: oil spills, dams, artefacts like wind mills, etc. In those cases we have direct moral responsibility. But also when we do not know what has caused their bad luck, it is our duty to prevent unnecessary harm to such animals. The main difficulty is to determine what counts as 'unnecessary' harm. The case of the lonely, screaming seal babies on the Dutch, German and Danish beaches demonstrate that people, who remove these animals, often fail to see that they have caused the mother to abandon her baby, because they have touched the baby the mother will not return. Also the stress involved in the 'rescue' attempt, and the future dependency on human care, have to be taken into account before bringing such an animal to a shelter. When animals are beyond help, the same first principle urges us to kill the animal in order to avoid unnecessary harm. This leads to the *second conclusion* about the killing of wild animals: *in a situation beyond help, either caused by bad luck or by human (indirect) doing, wild animals have to be killed, in a humane way*. This also is important in case of a life-threatening disease.

In specific eco-systems there are some animals that do not belong to the historical ecological and co-evolved biotic communities. They are member of, so-called, exotic species and are in danger of being totally eradicated by the management of these eco-systems. However, when we apply the third principle, the protection of the integrity or health of whole ecosystems (and of viable populations of the species contributing to that integrity or health), exotic species may be regarded of value in their own right insofar as they contribute to overall biodiversity. As a consequence the *third conclusion* is that *the total eradication of exotic species for all protected areas (no matter how impractical) has to be rejected*. Killing exotic species is only justified where there is proof of detrimental impacts on vulnerable native species that can be effectively alleviated (King, 1984). Scientific evidence that the historical biota has been disrupted, and that ecosystem health and biodiversity are declining, is not enough to motivate our intervention.

While the first and the second principle may be used to argue against some control programs, concern for animal welfare can also be used to argue for the need for such control. Some species are liable to reach high populations in the absence of any natural predation. Apart from the incidental danger to humans, this may cause major disruption to biotic communities, and welfare problems in circumstances of serious overpopulation (Koene and Gremmen, 2002). This, for example, has happened during severe winters to the deer population in the Oostvaardersplassen, a nature area near the city of Amsterdam in the Netherlands (Gremmen, 2014). After many debates the international oversight committee has decided to allow the killing of very 'weak' deer

(ICMO, 2012). When we generalize this decision, a *fourth conclusion* can be drawn: *in control programs the killing of very 'weak' animals is ethically justified on the basis that it is both good for ecosystems and also results in less aggregate harm and improved welfare for the populations.*

There are currently only two major means of control: hunting and (in) direct application of toxins. Where intervention is believed reluctantly to be unavoidably necessary, prima facie duties of non-interference to wild animals may have to be overridden. The *fifth conclusion* is that *justification is not always warranted for killing members of controlled species; only those necessary and effective for the purpose are ethically justified.* Even animal liberation ethicists and bio-spherical egalitarians, otherwise vehemently opposed to hunting, largely concede that truly necessary subsistence hunting is ethically justifiable (Taylor, 1986).

15.4.2 De-domesticated and feral animals

Although the concept of wildness is defined by freedom from human control, it admits to degrees insofar as there is greater or lesser human influence on the environment (Waller, 1998). In case of de-domesticated species humans were responsible for the dispersal of these species (Elliot, 1997). The ecological changes they bring about are ultimately anthropogenic. Recently introduced species, though immediately lacking historical associations in their new lands, just like any naturally invading species, may come to be regarded as naturalized once sufficiently co-adapted to their new environment (Woods and Moriarty, 2001).

The ethics of managing de-domesticated and feral animals has focused on duties to protect indigenous biodiversity, natural ecosystems, and animal welfare duties to minimize animal suffering. When we apply the first principle, taking the interests of animals seriously into ethical consideration and avoid unnecessary harm to sentient creatures, the *sixth conclusion* is that *one is morally obliged to care for de-domesticated animals;* for example, in the Dutch Oostvaardersplassen more than half of the Heck cattle and Konik horses have died of starvation in several severe winters (Gremmen, 2014). In line with our sixth conclusion, in such a situation one might be required to cull weak individuals rather than allowing them to die on their own. This has actually been the advice of the International Committee on the Management of large herbivores in the Oostvaardersplassen (ICMO) to the Dutch government, who has changed their management policy accordingly (ICMO, 2006). According to the deontological rights view we are responsible for what we do to animals in our care, but it is less clear that we have any duty to look after individual wild animals. Therefore even though a defender of the rights view may well object to the starting point of de-domestication, it is much less clear how he or she would consider later stages of the process.

15.4.3 Pest control

Especially feral animals may become a pest. Well known examples are rabbits in Australia, and mustangs in the USA. Ethical concerns about the destruction of free-living wildlife for disease control and environmental reasons have historically received little attention from animal welfare scientists, legislators or the public. For Littin *et al.* (2004: 3) pest control is very important because of the (in)direct impacts of pest species: 'in environmental degradation (e.g. soil erosion); endangerment of native plants and animals (e.g. by overgrazing, introduction of alien plant seeds, competition for food, predation); loss of primary production (e.g. by overgrazing and competition for food); and loss of livestock production through disease ... In addition, they also cause damage to private property and human health (e.g. damage to structures and electrical wiring, and transmission of tuberculosis)'.

In pest control animals are often killed – their welfare is deliberately compromised (i.e. it is deprived of life, and its death can be painful). A wide range of methods is used to kill or otherwise control unwanted wildlife. The animal welfare impacts of most of these methods are not known. Nevertheless, all vertebrates can experience pain and distress, regardless of whether they are unwanted pests or not. The capacity of a species to suffer in particular ways (e.g. pain compared to anxiety) may vary (Sommerville and Broom, 1998). The extremely high number of animals being controlled and the potential impacts of this control on their welfare suggest that killing them is often the only option. In view of the first and second principle the *seventh conclusion* is that *when animals become a pest, and therefore not only negatively impact the welfare of other animals, but often of the whole eco-system, they may be killed in the most humane way.* Even in those cases where a pest is just a temporally matter, pest control can be very important if. For example, in the case of a mice pest on farm lands the negative impact on the soil, the other animals and the loss of primary production could be very high.

15.5 Conclusions

In this chapter our approach has been to reformulate the three core questions of this book to the situation of wild animals. In answering the first question, what concepts are needed for the public and ethical evaluation of killing wild animals, we describe wildness as a broad concept, and equate it with parts of nature that are not controlled by humans. Our perspective on wildness is to consider it as a quality in specific individual animals, of being wild or un-wild. We differentiate between nine categories of animals in natural areas, and wild animals are considered to be at one end of a continuum and domesticated animals are at the other end. Thus, a development is possible from the wild stage to the pseudo-domestication stage and back again to the semi-wild stage. The description of an ethical framework of three principles enabled our affirmative answer to the second question, is it possible to justify the killing of a wild animal, and if so under what conditions. When we apply the ethical framework to the killing of wild animals, de-domesticated and feral animals, and to the killing of animals in pest control, the answer to the third question: 'Can we legitimately differentiate the issue of killing wild animals in different wild animal contexts', leads to seven conclusions.

These answers enable us to tackle the main question of this chapter about the ethics of killing wild animals: 'Will wild make a moral difference?' Compared to the killing of production animals, often kept with the sole purpose to kill them, the answer must be affirmative because already our first conclusion made it clear that we are not allowed to kill wild animals for our own purposes. Only in a situation beyond help, either caused by bad luck or by human (indirect) doing, wild animals have to be killed. In case of exotic species, controlled animals, de-domesticated and feral animals, and even when animals become a pest, only when intervention is believed to be unavoidably necessary, prima facie duties of non-interference to wild animals may have to be overridden. This means that in general the ethical policy call for wild animals is a 'hands-off' recommendation.

References

Beauchamp, T.L., 2003. The nature of applied ethics. In: Frey, R.G. and Wellman, C.H. (eds.) A companion to applied ethics. Blackwell Publishers, Malden, MA, USA, pp. 1-16.

Beauchamp, T.L. and Childress, J.F., 2001. Principles of biomedical ethics. 5th edition. Oxford University Press, New York, NY, USA.

Brophy, B., 1972. In pursuit of a fantasy. In: Godlovitch, S., Godlovitch, R. and Harris, J. (eds.) Animals, men and morals: an inquiry into the maltreatment of non-humans. Grove Press, New York, NY, USA, pp. 124-125.

Callicott, J.B., 1980. Animal liberation: a triangular affair. Environmental Ethics 2: 311-338.

Callicott, J.B., 1987. The conceptual foundations of the land ethic. In: Callicott, J.B. (ed.) Companion to a Sand County almanac: interpretive and critical essays. University of Wisconsin Press, Madison, WI, USA, pp. 186-220.

Callicott, J.B., 1989. In defense of the land ethic: essays in environmental philosophy. State University of New York Press, Albany, NY, USA, 325 pp

Eggleston, J.E., Rixecker, S.S. and Hickling, G.J., 2003. The role of ethics in the management of New Zealand's wild mammals. New Zealand Journal of Zoology 30: 361-376.

Elliot, R., 1997. Faking nature. The ethics of environmental restoration. Routledge, London, UK, 192 pp

Evanoff, R.J., 2005. Reconciling realism and constructivism in environmental ethics. Environmental Values 14: 61-81.

Gamborg, C., Gremmen, B., Christiansen S.B. and Sandoe, P., 2012. De-domestication: ethics at the intersection of landscape restoration and animal welfare. Environmental Values 19: 57-78.

Gremmen, B., 2014. Just fake it! Public understanding of ecological restoration. In: Oksanen, M. and Siipi, H. (eds.) The ethics of animal re-creation and modification. reviving, rewildering, restoring. Palgrave Macmillan, New York, NY, USA, pp. 134-150.

Hettinger, N. and Throop, B. 1999. Refocusing ecocentrism: de-emphasising stability and defending wildness. Environmental Ethics 21: 3-21.

Hickling, G.J., 1994. Animal welfare and vertebrate pest management: compromise or conflict? In: Baker, R.M., Mellor, D.J. and Nicol, A.M. (eds.) Animal welfare in the twenty-first century: ethical, educational and scientific challenges. Proceedings of the conference of the Australian and New Zealand Council for the Care of Animals in Research and Teaching (ANZCCART), April, Christchurch, New Zealand. ANZCCART, Glen Osmond, South Australia, pp. 119-124.

ICMO, 2006. Reconciling nature and human interests. Advice of the International Committee on the Management of large herbivores in the Oostvaardersplassen. Available at: http://www.rli.nl/sites/default/files/wintersterfteoostvaardersplassen6-2005adviesicmo.pdf.

Jamieson, D., 1995. Ecosystem health: some preventive medicine. Environmental Values 4: 333-344.

Klaver, I., Keulartz, J., Van den Belt, H. and Gremmen, B., 2002. Born to be wild: a pluralistic ethics concerning introduced large herbivors. Environmental Ethics 24: 3-21.

Koene, P. and Gremmen, B., 2002. Wildheid gewogen: samenspel van ethologie en ethiek bij de de-domesticatie van grote grazers. Wageningen University, Wageningen, the Netherlands.

Littin, K.E., Mellor, D.J., Warburton, B. and Eason, C.T., 2004. Animal welfare and ethical issues relevant to the humane control of vertebrate pests. New Zealand Veterinary Journal 52: 1-10.

Mepham, B., 1996. Ethical analysis of food biotechnologies: an evaluative framework. In: Mepham, B. (ed.) Food ethics. Routledge, London, UK, pp. 101-119.

Regan, T., 1983. The case for animal rights. Routledge and Kegan Paul, London, UK, 425 pp.

Rollin, B.E., 1996. Agricultural research and the new ethic for animals. In: Baker, R., Einstein, R. and Mellor, D.J. (eds.) Farm animals in biomedical research. Proceedings of the conference held in Wellington, New Zealand, 18-19 August 1995. ANZCCART, Glen Osmond, South Australia.

Rolston, H., 1988. Environmental ethics. Duties to and values in the natural world. Temple University Press, Philadelphia, PA, USA, 408 pp.

Rolston, H., 1994. Conserving natural value. Columbia University Press, New York, NY, USA, 259 pp.

Singer, P., 1990. Animal liberation. A new ethics for our treatment of animals. Avon Books, New York, NY, USA, 301 pp.

Sommerville, B.A. and Broom, D.M., 1998. Olfactory awareness. Applied Animal Behaviour Science 57: 269-286.

Staatsbosbeheer, 2012. Annual report 2012. Available at: http://tinyurl.com/puajhlv.

Taylor, P., 1986. Respect for nature. A theory of environmental ethics. Princeton University Press, Princeton, NJ, USA, 360 pp.

Vera, F.W.M., 2009. Large-scale nature development: the Oostvaardersplassen. British Wildlife 20: 28-36.

Waller, D.M., 1998. Getting back to the right nature. a reply to Cronon's the trouble with wilderness. In: Callicott, J.B. and Nelson, M.P. (eds.) The great new wilderness debate. University of Georgia Press, Athens, GA, USA, pp 540-567.

Woods, M. and Moriarty, P.V., 2001. Strangers in a strange land: the problem of exotic species. Environmental Values 10: 163-191

Index

A

B

C